Understanding the Role of Economic Operators within the Medical Device Regulatory Sector

Understanding the Role of Economic Operators within the Medical Device Regulatory Sector

Shalinee Naidoo

ARCLER
PRESS

www.arclerpress.com

Understanding the Role of Economic Operators within the Medical Device Regulatory Sector

Shalinee Naidoo

Arcler Press
224 Shoreacres Road
Burlington, ON L7L 2H2
Canada
www.arclerpress.com
Email: orders@arclereducation.com

ISBN: 978-1-77469-084-0 (Hardcover)

ABOUT THE AUTHOR

Shalinee Naidoo is currently the Regulatory and Product Development Manager of a medical device manufacturer based in South Africa. She is directly involved in global regulatory compliance and the design and development of new medical devices from idea conception to market. She is also the founder of Life of Shal (www.lifeofshal.com) an online travel journal created to inspire others to explore the world and Scientist's Sanctuary (www.scientistssanctuary.com) – a science communication company that specializes in bridging the gap between scientific knowledge and creative communication for both the academic and corporate world.

TABLE OF CONTENTS

LIST OF GLOSSARY

A

Active Implantable Device – Any medical device that relies on a source of electric energy or any other source of power other than that directly generated by the human body or gravity which is intended to be totally or partially introduced, surgically or medically, into the human body or by medical intervention into a natural orifice, and which is intended to remain after the procedure.

Adverse Event – Any event or problem that can or does result in permanent impairment, injury or death to the patient or the user of a medical device.

Audit – A systematic, independent examination of a Manufacturer's quality system that is performed at defined intervals to determine: (1) whether both quality system activities and the results of such activities comply with quality system procedures; (2) that these procedures are implemented effectively; and (3) that these procedures are suitable to achieve the set out quality system objectives.

B

Body Orifice – Any natural opening in the body. This includes the external surface of the eyeball or any permanent artificial opening.

C

CE Mark – symbol required for all medicines marketed into the European Union (EU) indicating that they meet common standards of performance and safety, known as essential requirements.

Clinical Data – Safety and/or performance information that are generated from the clinical use of a medical device.

Clinical Evaluation – The assessment and analysis of clinical data related to a medical device in order to verify the clinical safety and performance of the device when used as intended by the manufacturer.

Clinical Evidence – The clinical data and the clinical evaluation report pertaining to a medical device.

Clinical Evidence for an IVD Medical Device – All the information/data that supports the scientific validity and performance for its use as intended by the manufacturer.

Clinical Investigation – Any systematic investigation or study in or on one or more human subjects, undertaken to assess the safety and/or performance of a medical device. this term is similar in meaning to 'clinical trial' and 'clinical study.'

Clinical Trial – A clinical investigation or research study involving one or more human subjects which generates clinical data to be used to demonstrate the safety and effectiveness of a medical device.

Conformity Assessment – Systematic examination of evidence generated and procedures undertaken by a manufacturer, under requirements established by the Regulatory Authority. This is done to determine that a medical device is safe and performs as intended by the manufacturer and aligns or 'conforms' to the essential principles of safety and performance for medical devices.

Conformity Assessment Body (CAB) – A body, other than a Regulatory Authority, engaged in determining whether the relevant requirements in technical regulations or standards are met.

Contract Sterilizer – A person or business that provides sterilization services for another establishment's devices.

D

Distributor – May also include multiple distributors of the same device. This is a person or business that receives finished devices from another establishment for the purpose of offering them for commercial sale.

E

Effectiveness – A device is clinically effective when it produces the effect intended by the manufacturer relative to the medical conditions it is meant to treat.

H

Harm – Physical injury or damage to the health of people or damage to property or the environment.

Hazard – Potential source of harm.

I

Immediate Danger – A situation where the patient is at risk of either losing their life or an important physiological function if no immediate preventative measure is taken.

Implantable Device – Any device, including those that are partially or wholly absorbed and which is intended to be partially introduced into the human body through surgical intervention and intended to remain in place after the procedure for at least 30 days is also considered an implantable device.

Importer – A person or business that receives devices from foreign manufacturers for the purpose of offering them for commercial sale and/or distribution in the recipient country.

***In Vitro* Diagnostic (IVD) Medical Device** – this is a medical device intended by the manufacturer for the *in-vitro* examination of specimens derived from the human body solely or principally to provide information for diagnostic, monitoring or compatibility purposes.

Incident – An unusual or unexpected event that is associated with the use of a medical device and which may or may not lead to problems.

Indications for Use – A general description of the disease or condition the medical device or IVD medical device will diagnose, treat, prevent, cure, or mitigate, including a description of the patient population for which the medical device or IVD medical device is intended.

Instructions for Use – Information provided by the manufacturer to inform the user of the medical device of the intended purpose and proper use and of any precautions to be taken. This is also known as a "package insert."

Intended Use/Purpose – The objective or intent of the manufacturer regarding the use of a product, process or service as reflected in the specifications, instructions, and information provided by the manufacturer.

International Medical Device Regulators Forum (IMDRF) – A voluntary group of medical device regulators from around the world who have come together to build on the strong foundational work of the GHTF. This group aims to accelerate international medical device regulatory harmonization and convergence.

Invasive Device – A device which, in whole or in part, penetrates inside the body, either through a body orifice or through the surface of the body.

J

Jurisdiction – The power, right, or authority to interpret and apply regulatory laws.

L

Label – Written, printed, or graphic information either appearing on the medical device itself, or on the packaging of each unit, or on the packaging of multiple devices.

Labeling – The label, instructions for use, and any other information that is related to identification, technical description, intended purpose and proper use of the medical device but excluding shipping documents.

Lay User – an individual who does not have formal training in a relevant field or discipline.

Life Cycle – All phases in the life of a medical device, from the initial conception to final decommissioning and disposal.

Life Supporting or Life Sustaining – A device that is essential to, or that yields information that is essential to, the restoration or continuation of a bodily function important to the continuation of human life.

Local Manufacturer – A manufacturer within the country which the medical device is being sold.

Long Term – Normally intended for continuous use for more than 30 days.

M

Manufacturer – A person or business that produces an article that meets the legal definition of a medical device.

Market Surveillance – activities carried out and measures taken by competent authorities to check and ensure that medical devices comply with the requirements of Regulations set out.

Medical Device – Any instrument, apparatus, implement, machine, appliance, implant, reagent for *in vitro* use, software, material or other similar or related article intended by the manufacturer to be used, alone or in combination for human beings, for one or more medical purposes.

Medical Products – Medicines, vaccines, diagnostics, and medical devices.

P

Patient – An individual under the care of a healthcare provider who may benefit from the action of a medical device.

Performance – The ability of a medical device to achieve its intended purpose as stated by the manufacturer.

Placing On-Market Control – All activities that performed by local establishment (Importer/ distributor/Authorized Representative, etc.), such that it complies with regulatory requirements to ensure safety and effectiveness of medical devices.

Post Market Surveillance – Carried out systematically and proactively by the manufacturer to collect and review information/experience gained from their medical devices such that they ensure continued safety and efficacy.

Post-Market – When medical devices/products are already on the market.

Post-Market Control – All activities that are performed by all parties (either manufacturers or local establishments) such that they comply with regulatory requirements to ensure continued safety and effectiveness of medical devices during their use.

Problem – possible faults with the medical device, difficulties in using the device, or any undesirable outcome associated with the use of the device. A problem may not lead to an adverse event, but corrective or preventive actions are required.

Q

Quality Management System – The organizational structure, responsibilities, procedures, processes, and resources for implementing quality management.

R

Refurbisher/Remarketer – A person or business that assumes ownership of previously-used devices and reconditions them for resale.

Regulatory Authority (RA) – A government body or other entity that exercises a legal right to control the use or sale of medical devices within its jurisdiction, and that may take action to ensure that medical products marketed within its jurisdiction comply with legal requirements.

Relabeler – A person or business that changes the content of the product labeling offered by the original manufacturer for distribution under its own name.

Repackager – A person or business that packages finished devices from bulk or repackages devices made for shipment in multiple containers.

Reusable Surgical Instrument – Instrument intended for surgical use by cutting, drilling, sawing, scratching, scraping, clamping, retracting, clipping or other surgical procedures, without connection to any active medical device and which are intended by the manufacturer to be reused after appropriate procedures for cleaning and/or sterilization have been carried out.

Risk – The combination of the probability of occurrence of harm and the severity of that harm.

Risk Assessment – Overall process comprising a risk analysis and a risk evaluation.

Risk Evaluation – Procedure based on the risk analysis to determine whether tolerable risk has been exceeded.

S

Safety – Free from unacceptable risk.

Shelf-Life – the period of time until the expiry date during which a medical device in its original packaging maintains its stability under the storage conditions specified by the manufacturer.

Short Term – Normally intended for continuous use for not more than 30 days.

Stability – The ability of a medical device and IVD medical device to maintain its safety and performance characteristics within the manufacturer's specifications over a specified period of time.

Summary Technical Documentation (STED) – A summary of technical documentation held or submitted for conformity assessment purposes.

Surgically Invasive Device – An invasive device which penetrates inside the body through the surface of the body, with the aid or in the context of a surgical operation.

T

Technical Documentation – Documented evidence which is normally an output of the quality management system and which demonstrates compliance of a device to the Essential Principles of Safety and Performance of Medical Devices.

Transient – Normally intended for continuous use for less than 60 minutes.

U

User – The person (professional, lay, or patient) who uses a medical device.

V

Vendor – Any person who sells medical devices, such as a manufacturer, an importer, a distributor, a wholesaler, or a retailer.

Vigilance – Reporting of serious incidents and field safety corrective actions by manufacturers to the relevant competent authorities.

LIST OF FIGURES

LIST OF TABLES

LIST OF ABBREVIATIONS

AHWP	Asian Harmonization Working Party
AIMD	Active Implantable Medical Devices
AIMDD	Active Implantable Medical Device Directive
AR	Authorized Representatives
ASR	Articular Surface Replacement
BSI	British Standards Institution
CA	Competent Authority
CEN	Committee for Standardization
DHF	Design History File
DoC	Declaration of Conformity
EEA	European Economic Area
EP	Essential Principles
EU	European Union
EUDAMED	European Database for Medical Devices
FDA	Food and Drug Administration
FSCA	Field Safety Corrective Actions
GDP	Good Distribution Practices
GHTF	Global Harmonization Task Force
GMDN	Global Medical Device Nomenclature
GMP	Good Manufacturing Practices
GSPR	General Safety and Performance Requirements
ICIJ	International Consortium of Investigative Journalists
IFU	Instructions for Use
IMDRF	International Medical Device Regulators Forum
ISO	International Organization for Standardization
IUDs	Intrauterine Devices
IVD	*In Vitro* Diagnostic

IVDD	*In Vitro* Diagnostic Medical Devices
IVDMDD	*In Vitro* Diagnostic Medical Device Directive
IVDR	*In Vitro* Diagnostic Regulation
MDA	Medical Device Alert
MDCG	Medical Device Coordination Group
MDD	Medical Device Directive
MDR	Medical Device Regulation
MDSAP	Medical Devices Single Audit Programs
MHRA	Medicines and Healthcare products Regulatory Agency
MoM	Metal-on-Metal
MRI	Magnetic Resonance Imaging
NB	Notified Bodies
NHS	National Health Service
NJR	National Joint Registry
PIP	Poly Implant Prothese
PRRC	Person Responsible for Regulatory Compliance
QMS	Quality Management System
STED	Summary Technical Documentation
UDI	Unique Device Identifier
WHO	World Health Organization

PREFACE

Access to good quality and affordable health care products is essential to ensuring global advancements in health coverage while at the same time attempting to advance solutions to health-related emergencies. Medical devices are becoming more important in the health care sector, and this important role increases the need for the overall provision of safe, efficient, and regulated devices. For countries to achieve their health-related development goals, as well as to ensure meeting of the Millennium Development Goals; it is various imperative disciplines related to medical devices such as manufacturing, regulatory, planning, design, assessment, management, and designation of use, all lead to the production of good quality, safe, and compatibility devices, that function appropriately in the settings which they are meant to be used. In recent times there has been an unprecedented level of public interest regarding the regulation of medical devices in Europe due to the strict safety requirements of the European Union (EU). The release of the new medical device regulation (EU MDR 2017/745) in the Official Journal of the European Union in May of 2017 marked the start of a three-year transition period for manufacturers, suppliers, notified bodies (NB), and national Competent Authorities. Compliance with the new regulation was set to take effect in May of 2020 but has now been postponed by a year due to the coronavirus pandemic. The scale of the changes required to be made is significant and has put a great deal of pressure on manufacturers. Often, it is a common misconception that only the manufacturer is responsible for ensuring compliance; however, it is, in fact, the responsibility of the manufacturer as well as the distributor, importer, authorized representative, and notified body. This volume aims to provide an easy-to-understand guide to changes between the current medical device directive (MDD) 93/42/EEC and forthcoming Medical Devices Regulation (EU MDR 2017/745) with regards to classification, conformity assessments, and assessing general safety and performance requirements (GSPR) as well as the roles and responsibilities of designated economic operators.

Chapter 1

The History of Medical Device Regulations

CONTENTS

1.1. THE HISTORY OF MEDICAL DEVICE REGULATIONS (MDRS)

Prior to the 1930s, the medical device industry was practically non-existent. Much of the medical devices being manufactured were sold directly to the public without any form of regulations to guide performance, safety, or functionality. After World War II, the medical device industry and related technologies grew exponentially, leading to the first developments of any kind of regulating between industry, government, and the public. By the late 1960s, thousands of medical device manufacturers began to ship their devices. In fact, it was estimated that the total shipped products amounted to a value exceeding $1 billion. In the early 1970s, a released government report documented more than 10,000 injuries as a result of medical devices. Medical device regulations (MDRs) tend to vary a great deal from country to country. Over the last 20 or so years, the number, range, and complexity of medical devices have increased and as a result, concurrently, so has the complexity of the regulations involved. The last twenty years or so have seen an explosion in the variety, use, and complexity of medical devices. These devices vary greatly in complexity, ranging from implantable devices such as pacemakers to less risky devices such as bedpans and gloves. In 2016 alone, the total global medical device industry was estimated at $130 billion, with almost half of the global production and consumption occurring in the United States alone.

In 2001, the WHO published an "A Model Regulatory Program for Medical Devices," which served as an international guide to assist member states in establishing regulatory programs for medical devices. The aim of this model was to provide information to nations that lacked a medical device regulatory system such that this guide would then enable the production of devices that met internationally compatible regulations. This was then followed by the release of "Medical Devices Regulations: Global Overview and Guiding Principles" in 2003 by WHO. The main purpose of this guide was to emphasize the complexity of the medical device industry and identify issues related to regulations. This document essentially provides guidance to member states who want to create new or modify existing medical device regulatory systems. By 2010, the WHO had then released an estimated report detailing an overview of countries and related medical device regulatory frameworks (Table 1.1) numerically.

Table 1.1: Overview of Country Percentages and Related Medical Device Regulatory Frameworks

Countries (%)	Medical Device Regulations
30	Countries that have a developed framework for the regulation of medical devices.
30	Countries that partially have regulations of medical devices.
40	Countries that are either developing a framework or do not yet have any regulation.

1.2. INCIDENTS THAT INFLUENCED THE REGULATORY SECTOR FOR MEDICAL DEVICES

All around the world, medical devices have been involved in various negative incidents on patients, which have either injured and, in worse cases, led to patient death. On the other hand, health authorities and various regulatory bodies across the globe have subsequently failed to protect such patients from these poorly functioning (and tested) devices, which have had various negative effects on patients. These include the puncturing of organs, erroneous shocks, rotting of bones and organs, blood poisoning, and other effects. A year-long investigation by the International Consortium of Investigative Journalists (ICIJ) in 2018 revealed that many governments actually hold many of these implants (regardless of the complexity or risk classification) to much lower safety testing standards as when compared to pharmaceuticals. Flaws within the regulatory field itself have resulted in medical device manufacturers leaving such problematic devices on the market in some countries while pulling them off the market in other countries. In fact, the Implant Files Investigation conducted by ICIJ, which formed one of the biggest and first-ever global examinations of the medical device industry, revealed just how health and regulatory authorities all around the world have failed to protect millions of patients due to them being uninformed of the crucial risks around such devices. Many of these reported incidents and various past ones have highlighted the need for improvement across standards, processes, procedures, and overall legislation where medical devices are concerned. Many of these incidents have directly highlighted the need for strengthening existing legislation, specifically within the European Union (EU), which has, in turn, led to drastic improvements and changes over the past decade, ultimately giving regulators more control and oversight.

Previous European legislation adopted in the 1990s aimed at ensuring the smooth functioning of the internal market together with an attempt to maintain high health and safety levels. There have, however, been several key moments in an industry that have highlighted the shortcomings of such legislative frameworks.

In April 2017, the new EU Medical Devices regulations and IVD regulations were adopted by the European Parliament. These new regulations entered into force on the 25th May 2017 and stated that the fundamental revision of the legislation was to provide a 'sustainable regulatory framework for medical devices which ensure a high level of safety and health while supporting innovation.' The main purpose of the changed regulations was to ensure smooth functioning of the internal market with regards to medical devices and enforce a high level of health protection for patients and users.

1.3. THE PIP BREAST IMPLANT SCANDAL

The PIP breast implant scandal was one of the biggest medical device industry scandals that ultimately influenced and resulted in major regulatory updates, not only within the EU but on a global level. Medical devices in the EU are regulated by the Medical Devices Directive, which had also been transposed in France in several provisions of the French Public Health Code and in England in the Medical Devices Regulations (2002).

In 1965, plastic surgeon Henri Arion introduced breast implants to France. Together with Jean-Claude Mas, they launched a Poly Implant Prothese (PIP) company in 1991. This company essentially produced a figure of around two million sets of silicone breast implants over a 20-year period; however, despite their success, they also created one of the biggest global health scares. This changed the medical device regulatory field.

The health scare came about when PIP decided to change the material being used to manufacture the silicone breast implants to a cheaper, industrial-grade silicone that was not approved for medical use. Industrial-grade silicone is thought to have more contaminants than medical-grade silicone; however, it is thought the company chose to use the alternative to save costs and boost overall profits. Investigations later revealed that this industrial-grade silicone had a rupture rate that was double that of the general industry average. In addition, later studies revealed that if compromised, the silicone gel was thought to cause both short, and long-term effects such as inflammation and possible scarring in affected patients.

In December 2011, French authorities issued a worldwide alert, advising all women who had been fitted with potentially defective implants to have them removed. By this time, PIP had become the world's third-largest supplier of implants, with over 300,000 women in 65 different countries around the world have received the implants, be they for cosmetic reasons or medical reasons such as reconstructive surgery after breast cancer treatment. Much of the implants had been used widely across Europe. It is estimated that around 40,000 women in Britain and South America alone may have received such implants.

Much of the advice given to women from regulatory bodies and surgeons alike varied from country to country. In France, women who had received such implants were urged to have them removed as a precaution. In Venezuela, regulatory authorities further recommended removal while the British government, on the other hand, said there was no evidence to recommend routine removal of PIP implants. They did, however, advise women who were concerned to have them removed. In Germany, medical groups advised women to seek removal of their implants with a further statement indicating that there was no immediate danger.

The French government subsequently offered to pay for the removal and replacement of all PIP implants fitted in France, while Venezuela said it would cover the cost of implant removals but not the replacements. Within the United Kingdom, those women who had their implants paid for by the National Health Service (NHS) was able to have them removed free of charge after a consultation with their doctor, while private companies who had fitted PIP implants for cosmetic purposes were advised to follow suit.

1.4. OVERVIEW OF EVENTS IN THE PIP SCANDAL (TABLE 1.2)

Table 1.2: Overview of the Events in the PIP Scandal

1991	PIP was launched by Jean-Claude Mas
1992	The United States Food and Drug Administration (FDA) calls for a voluntary moratorium on the use of silicone gel Implants. Implant sales are halted in the United States.
1994	PIP begins selling hydrogel implants.
1996	PIP begins selling saline-filled implants in the United States.
1997	PIP authorized to produce medical-grade silicone implants.
2000	The FDA refuses to approve PIP's saline-filled implants and warns about deviations from good manufacturing practices (GMP) found at the PIP plant. The company withdraws its hydrogel implants from the market as it cannot show they are safe.
2001	PIP starts using an unapproved industrial-grade silicone in their implants.
2003	Regulatory bodies begin to trace the first signs of legal problems and financial losses.
2009	Surgeons started reporting abnormally high rupture rates in France, sparking concern. This led to a host of legal complaints and PIP undergoing bankruptcy. The UK regulatory authority subsequently warned of several medical claims being made.
2010	PIP was placed into liquidation after the French medical safety agency recalled its implants.
2011	The French government recommended that approximately 30,000 women with PIP implants seek removal of the implants as a health precaution.
2012	Jean-Claude Mas, the founder of PIP is arrested. France launches an investigation into the scandal.
2013	Jean-Claude Mas is sent to prison for four years and fined 75,000 euros. He is subsequently banned for life from working in medical services or running a company. He also faced two additional legal cases: one for involuntary manslaughter and another linked to the financial implications of the scandal.
2016	Mas's sentence upheld by the appeal court.
2017	New European Medical Devices Regulation (MDR) was published and in force, with a transition period for manufacturers to comply over 3 years.

1.5. WHAT HAPPENED TO PIP?

PIP implants were banned, and eventually the company went bankrupt in 2010. The founder of PIP, Jean Claude Mas went under investigation by the French police and was subsequently charged with involuntary injury over the implants. PIP's lawyer eventually did respond to the scandal stating that industrial silicone was used in their implants and that while no tests conducted showed any danger, the main issue was public fear which subsequently created intense media pressure and pushed the blame ultimately onto PIP. Regardless, the concerns raised over the PIP implants have led to one of the biggest calls for changes to regulations within the plastic surgery industry. Ultimately the consumer does not know what should be regarded as being safe or not hence the need for more stringent regulations.

The events of the PIP scandal resulted in the European Commission seriously reviewing and changing their MDRs. In fact, the regulation of the medical device industry was already in the process of being reformed when the PIP scandal occurred. The scandal however increased the focus and need for such change. In 2008, the European Commission published a public consultation document that asked stakeholders for their view around the revision of the legal framework of medical devices. This was followed by a public consultation in 2010 which focused on the technical aspects related to the revision of Directive 98/79/EC on *in vitro* diagnostic (IVD) medical devices.

1.6. CHANGES TO THE MEDICAL DEVICE REGULATIONS (MDRS) AS A RESULT OF PIP

Ultimately the main reasons for such an industry overall were that, at the time, existing EU regulations, which dated back to the 1990's had not kept pace with the technological progression over the years. This had resulted in many regulatory gaps and uncertainties around products being manufactured from non-viable human tissues or cells as well as implantable and/or other invasive products used for cosmetic purposes. Furthermore, there were substantial divergences in the overall interpretation and application of the regulations/rules that emerged between EU countries. Much of this varied interpretation arose from each country having to interpret and respond to such technological and scientific progression individually. In addition, there was limited ability to trace medical devices back to their suppliers. Essentially the new regulations, now referred to as the European Medical Devices Regulation (MDR), which also includes regulations (EU) 2017/745 and 2017/746 have become far more stringent. Some of the major changes include:

- Expansion of the overall scope within the MDR to cosmetic and/or esthetic devices.
- Clarification regarding areas such as medical software and generic tests.
- Reflection of technological advances through the adaptation of safety and performance requirements that are applicable to new health technologies.
- Stricter requirements for medical device compliance from manufacturers to ensure overall patient safety.
- Stronger supervision of independent assessment bodies by national authorities.
- More power and obligations for those above-mentioned assessment bodies to ensure thorough testing and regular checks on manufacturers such as unannounced factory inspections, sample testing.
- Improvement of co-ordination between national surveillance bodies.
- The creation of a national registry to increase traceability. This allows manufacturers to register themselves and the devices they place on the EU market within a central database through a unique device identification system. This way, it will make it easier to trace and identify devices and enhance overall post-market safety.
- A ban on aggressive marketing of medical devices.
- Overall requirements for improved training and qualification standards.
- Increased and improved transparency along the entire supply chain of medical devices.
- The establishment of a portal within the EU, which requires manufacturers to report serious events and corrective actions to reduce the overall risk of such events happening again.
- An increase in the rights and responsibilities of notified bodies (NB) as well as the stringency with which they monitor all medical devices.

- A further requirement that within a manufacturers organization to have a qualified person responsible for regulatory compliance (PRRC).
- Patients that have been implanted with a device should also be given essential information which will allow the device to be identified. This information should also contain necessary warnings or precautions that must be taken.

1.7. THE DEPUY HIP REPLACEMENT RECALL

MoM hip designs were first introduced in the 1950s and 1960s until such time that improvements in manufacturing and measurement technologies sparked designers to revisit the overall design of MoM hips in the 1990s. In 2003, the medical implant manufacturer DuPuy Orthopedics which is a subsidiary of Johnson & Johnson, introduced their own version of a metal-on-metal (MoM) hip resurfacing articular surface replacement (ASR) ASR™ and ASR™ XL, a total hip replacement version. Both versions were available in Europe and were available under the clause of 'substantial equivalence' which, allows for fast-track market accreditation through claiming the implant in question is similar to ones already on the market. As per regulatory bodies, the ASR™ hip resurfacing was granted market approval by the FDA in the USA while the ASR™ XL was available globally.

In 2008, problems with ASR™ were first raised at various orthopedic conferences as well as in journal publications. This was followed by an annual report published by the National Joint Registry (NJR) of England and Wales in 2009, which showed high revision rates of approximately 7.5% at three years for the ASR™ hip resurfacing. At the time, however, official notices pertaining to such problems were limited to guidance on positioning during surgery until around August 2010 when DuPuy Orthopedics issued a global recall of all ASR™ and ASR™ XL hip systems. In addition, the Medicines and Healthcare products Regulatory Agency (MHRA) further issued a general medical device alert (MDA) on 22 April 2010 for all metal-on-metal (MoM) hip replacements (MDA/2010/033). This MDA did not specifically address the ASR™ implants but rather showed that much data had been available for a while, indicating that general concerns around MoM hip implants did exist. It should also be noted that prior to this, high revision rates with ASR™ were also highlighted in Australia somewhere in 2007. As a consequence of this, the ASR™ was withdrawn at the end of 2009.

1.8. OVERVIEW OF EVENTS IN THE DUPUY SCANDAL

The impact of such high revision rates on patients is significant and should be noted. Failed MoM hip implants can cause both local and systemic health problems as well as symptoms such as hearing loss, dizziness, decline in cognitive functions, cardiomyopathy, and organ failure while ultimately could result in increased pain and a decrease in overall mobility. Various reasons were attributed to the failure of the hip replacement system. These include loosening and misalignment of components, infection, fracture of the bone, dislocation, metal sensitivity, and overall pain. In addition, further complications could include an increase in the level of metal ions in the blood, staining of the bone, necrosis, swelling, and unknown damage to the nerves, tissues, and/or muscles (Table 1.3).

Table 1.3: Overview of Events in the DuPuy Scandal

Overview of Events in the DuPuy Scandal	
2003	ASR™ resurfacing introduced
2007	High revision rates reported by the Australian Orthopedic Association, National Joint Replacement Registry
2008	Subsequent reporting of high revision rates by the Australian Orthopedic Association, National Joint Replacement Registry
2009	First report of high revision rates in National Joint Registry of England and Wales ASR™ withdrawn in Australia and New Zealand
2010	MHRA issues MDA on all MoM implants noting a small number of patients have adverse reactions DePuy release guidance on positioning MHRA issue MDA on positioning DePuy withdraw the ASR™ globally

1.9. REACTIONS TO PUBLISHED DATA ON DUPUY HIP REPLACEMENT SCANDAL

In 2007, the Australian Orthopedic Association national joint replacement registry noted that ASR seemed to have twice the risk of revision, thus indicating the need for further surgery as when compared with other resurfacing procedures. This was followed by the registry publishing data the next year, showing that some women undergoing MoM hip resurfacing were facing more follow-up operations.

At the beginning of 2010, DuPuy Orthopedics issued a statement saying they were phasing out the ASR Hip Implant due to declining sales; however, they failed to mention the data from the Australian implant registry. Later, sometime in March 2010, DuPuy issued their first warning to doctors and patients indicating the high early failure rate but still had not issued a product recall. Eventually, DuPuy Orthopedics withdrew their implants from the market after the NJR in England and Wales shared data that showed patients with their MoM hips having a higher expected revision rate. In fact, data showed that within a five-year period of receiving DuPuy's resurfacing device, 12% of patients had undergone a revision surgery, and within five years of having an ASR total hip replacement, 13% of patients had undergone revision surgery.

It was estimated that about 93,000 people globally had received a DuPuy implant. The first lawsuit against DuPuy Orthopedics was filed in the United States on June 15, 2010, followed another on September 27, 2011. Both lawsuits claimed that DePuy ASR hip replacements were improperly designed and that DePuy knew there were problems with their implants early on and, despite this, did not alert patients or surgeons about the possible problems. In 2014, Johnson & Johnson announced that it would not withdraw a $2.5 billion global settlement. Settlements began at a base of $250,000. Johnson & Johnson continued to deny culpability. It should be further noted that of the total number of people known to receive the implants, approximately 4,700 were thought to be from India. Johnson & Johnson then undertook to working with the Indian government in 2018 in an attempt to support all Indian patients affected by the implants.

1.10. INADEQUATE VAGINAL MESH

Back in the late 1980s, the medical industry was searching for new ways to treat women that were experiencing urinary incontinence as well as vaginal prolapse. Both are common conditions that women experience after childbirth and while most doctors, at the time, suggested physiotherapy, weight-loss, and other non-surgical interventions as a solution; the use of mesh implants seemed like a simple, quick, and convenient alternative. The mesh was initially designed to allow a patient's body tissues to grow through it, making it harder to remove. Mesh implants took about an hour to implant and allowed women to leave the hospital quickly and soon became the standard treatment for millions of women all around the world. While many cases have had effective results, there are also some women who have experienced severe complications such as eroding mesh that has pierced the vaginal wall or bladder, nerve damage, and infections. In addition, many of these women have experienced chronic pain and, in some cases, were barely able to walk.

One of the largest manufacturers of gynecological mesh is medical device manufacturer, Boston Scientific. While the company has been successful in the past, they have also attracted close to 48,000 lawsuits against them, all of which claim that the gynecological mesh used can cause extreme pain and injury. Despite such claims, Boston Scientific has continued to fight such allegations with a comeback stating that over one million women have been successfully treated and that they have extensively tested the plastic/resin used to confirm the mesh composition, safety, and performance. Unfortunately, many doctors have also found otherwise, with some claiming the mesh causes a chronic inflammatory reaction and stating that once removed, the mesh seems to have a substantially altered architecture. In addition, they seem to have shrunk by 50% in width and are excised in scar tissue.

Investigations revealed that the mesh is made of a polypropylene plastic which is known to be a common packaging material. Boston Scientific however has proof that they have clearance from the FDA to use a particular brand of polypropylene called Marlex. This brand is said to be made in Texas by a subsidiary of Chevron Phillips. In 2004 however, Chevron Phillips became concerned with the use of Marlex in the medical industry and issued a warning that it must not be used for any form of permanent implantation in the body. This is due to the unstable oxidative state of polypropylene. Oxygen is known to break up the polypropylene and cause dissipation of antioxidant additives over time and as a result, causes the polypropylene to fall apart within the body. As time passed, Chevron Phillips cut off Boston Scientific's supply of Marlex, somewhere in 2005; however, Boston Scientific remained adamant on continuing with their processes. As a result, they continued their search for a supplier and eventually came across a broker in China who claimed to have large amounts of Marlex that was supposedly imported from Chevron Phillips in Texas. Simultaneously that same month, the FDA issued a report claiming that over a five-year period, they had found about 4000 reports of injury, death, and malfunction as well as associated complications including pain, infection, urinary problems, bleeding, and organ perforation. This further gave the industry a reason to believe that if they planned on switching plastics, the FDA would now require years of testing, which ultimately might fail and hinder sales.

The Chinese company seemed like the best option at the time. Boston Scientific had established their own procedures for what required documents would be needed as well as requests for various import records to prove the Marlex was really imported from Texas by the Chinese company. Unfortunately, the Chinese company did not have any documents that could ensure the authenticity of the Marlex could be verified. Further investigation revealed that lot numbers on the bags did not match up as well as counterfeit printing. Boston Scientific then further ordered testing on the material and analyzed 11 different parameters. Results from nine of these tests differed. Despite this, Boston Scientific still bought a large amount of Chinese plastic and continued to use it in their manufacturing processes.

Chapter 2

The CE Mark Process

CONTENTS

Medical devices are important to providing adequate health care and to improving the health of individuals all around the world. Without properly regulated medical devices, many routine medical procedures, whether they be bandaging an open wound or implanting an artificial hip, would be impossible. As innovation and rapid advancement in technologies take over, medical devices have now become one of the fastest-growing industries globally. In fact, in 2000, it was estimated that one and half million different medical devices were available on the market worldwide.

Regulatory controls for medical devices are often complicated to implement due to the many different types of devices available on the market. Often the risks associated with each type of device differ, making the opportunity for misuse considerable. For these reasons, it is not justifiable to subject all types of devices to the highest level of regulatory control, and many countries tend to have adopted controls from other countries. Advancing technology is producing an overwhelming number of new devices onto the market, and as such, one of the biggest issues affecting the medical device industry today is the diversity of devices available. In recent times there has been an unprecedented level of public interest regarding the regulation of medical devices in Europe. This is because devices cannot be placed on the European market without conforming to the strictest of safety requirements of the European Union (EU).

In order to sell medical devices within the EU, medical device manufacturers will need a valid MDD (medical device directive) certificate. Collectively known as the MDD, this core legal framework consists of three directives that regulate the safety and marketing of medical devices in Europe. These directives came into effect in the 1990s and are as follows:

- Active implantable medical device directive (AIMDD 90/385/EE);
- Medical device directive (MDD 93/42/EEC);
- *In vitro* diagnostic medical device directive (IVDMDD 98/79/EC).

Due to the new technologies, increasing incidents within the medical device industry and more apparent gaps in current framework there was increasing need to update the MDD which led to the development and release of the medical device regulation (EU MDR 2017/745). Under the new EU medical device regulation (2017/745), manufacturers will be assessed according to these new regulations and will thus require new certificates for their devices and products. Companies wishing to continue selling their medical devices and products in the EU after enforcement of the MDR will need to transition their technical product documentation as well as the documentation within their quality management system (QMS). Many manufacturers are unaware of how long the transition to the new MDR regulation actually takes, and for this reason, ensuring a thorough understanding of what the requirements are asking is imperative for success. If manufacturers do not start planning early enough, it could lead to no valid certificate being issued for a long period of time which in turn may result in businesses not being able to access the European market. Non-compliant notified bodies (NB), similarly, may lose re-designation. Since the MDR's release in 2017, many regulatory experts have published and circulated resources aimed at helping companies understand the new rules and adjust accordingly. This volume aims to do the same but on a much simpler level, offering easy to understand explanations of the major regulatory transitions and changes associated with Articles 1 to 15 of the new medical device regulation (MDR) 2017/745.

2.1. REGULATION OF MEDICAL DEVICES IN THE EUROPEAN UNION (EU)

Medical devices cannot be placed on the European market without a manufacturer proving conformance to the strict safety requirements of the EU which, once done, allows for the affixation of the CE conformity mark on the device itself. These safety requirements are laid out in relevant regulations, previously known as the medical device directive (MDD) and now the MDR. The MDD formed the core legal framework in Europe since the 1990s. it consists of three directives that regulate the safety and marketing of medical devices. These three directives are:

- Active implantable medical device directive (AIMDD 90/385/EE);
- Medical device directive (MDD 93/42/EEC); and
- *In vitro* diagnostic medical device directive (IVDMDD 98/79/EC).

These have been supplemented since by several necessary updates and relevant guidance documents and directives to guide manufacturers when implementing the regulation. Often these guidance documents and directives are released due to new and emerging technologies which challenge current frameworks or highlight gaps or areas of confusion.

The MDD defines a Medical Device as:

'Any instrument, apparatus, appliance, material or other article, whether used alone or in combination, including software necessary for its proper application intended by the manufacturer to be used for human beings for the purpose of:

- Diagnosis, prevention, monitoring, treatment or alleviation of disease;
- Diagnosis, monitoring, treatment, alleviation of or compensation for an injury or disability;
- Investigation, replacement or modification of the anatomy or of a physiological process;
- Control of conception.

and which does not achieve its principal intended action in or on the human body by pharmacological, immunological or metabolic means, but which may be assisted in its function by such means.'

2.2. UNDERSTANDING THE CE MARKING PROCESS

The CE mark is a conformity mark which all European medical devices must have before they can be placed on the European market. By placing this mark on a device, it is seen as a declaration by the manufacturer that the product meets all the appropriate provisions of the relevant legislation, including those related to safety and, where required, has been assessed in accordance with these.

Article 17 (CE Marking) of the MDD 93/42/EEC, States:
1. 'Devices, other than devices which are custom-made or intended for clinical investigations, considered to meet the essential requirements referred to in Article 3, must bear the CE marking of conformity when they are placed on the market. 2. The CE marking of conformity, as shown in Annex XII, must appear in a visible, legible, and indelible form on the device or its sterile pack, where practicable and appropriate, and on the instructions for use. Where applicable, the CE marking must also appear on the sales packaging. It shall be accompanied by the identification number of the notified body responsible for implementation of the procedures set out in Annexes II, IV, V, and VI. 3. It is prohibited to affix marks or inscriptions which are likely to mislead third parties with regard to the meaning or the graphics of the CE marking. Any other mark may be affixed to the device, to the packaging or to the instruction leaflet accompanying the device, provided that the visibility and legibility of the CE marking is not thereby reduced.

The CE conformity mark consists of the initials 'CE' and is usually represented as in the figure below.

The letters "CE" are the abbreviation of the French phrase "Conformité Europééne" which literally translates into the phrase "European Conformity." The term initially used was "EC Mark" and it was officially replaced by "CE Marking" in the Directive 93/68/EEC in 1993. "CE Marking" is now used in all EU official documents. CE Marking on a product is a manufacturer's declaration that the product complies with the essential requirements of the relevant European health, safety, and environmental protection legislation and Directives. These directives contain the "essential requirements" and/or "performance levels" and "Harmonized Standards" to which the products must conform. The Harmonized Standards are the technical specifications (European Standards or Harmonization Documents) which are established by several European standards agencies (CEN, CENELEC, etc.).

- CEN stands for European Committee for Standardization.
- CENELEC stands for European Committee for Electrotechnical Standardization.

CE Marking on a product indicates to governmental officials that the product may be legally placed on the market in their country. It ensures the free movement of the product within the EU and permits the withdrawal of the non-conforming products by customs and enforcement/vigilance authorities.

2.3. GENERAL PRINCIPLES OF THE CE MARK

There are specific requirements for the visual display of the CE mark-these include that the various components of the CE mark must have the same vertical dimension, which may not be less than 5 mm (although this minimum dimension may be waived for small-scale devices). It is also important to understand that the CE mark or rather the letters "CE" do not stand for any specific words, but rather it is a symbol indicating a declaration by the manufacturer that the product meets all the appropriate provisions of the relevant legislation. It should not be placed on a device if the manufacturer cannot prove conformity to the regulation and that the device has been assessed in accordance with the appropriate procedures. The CE mark also means that the product can be freely marketed anywhere in the European Economic Area (EEA) without further control.

The CE marking shall be affixed only by the manufacturer or his authorized representative and should be affixed only to those products to which its affixing is provided for by specific harmonized legislation, and shall not be affixed to any other product. By affixing or having affixed the CE marking, the manufacturer indicates that he takes responsibility for the conformity of the product with all applicable requirements set out in the relevant legislation.

Any other marking may be affixed to the product provided that the visibility, legibility, and meaning of the CE marking is not impaired. Member States shall ensure the correct implementation of regulatory bodies governing the CE mark and shall take appropriate action in the event of improper use of the marking. Member States shall also provide for penalties around infringement of the CE mark. These penalties shall be proportionate to the seriousness of the offense and shall constitute an effective warning against improper use.

2.4. REGULATORY BODIES INVOLVED IN THE CE MARKING PROCESS

There are a range of regulatory bodies involved in the CE marking process. These include Competent Authorities (CA), NB, and the authorized representatives (AR). It is important for a manufacturer to understand the difference between each one. These are explained below:

2.4.1. Competent Authorities (CA)

A competent authority (CA) is defined as a regulatory body authorized by a European Member State government to monitor compliance of devices with the national legislation or regulations and carry out duties on behalf of the government in compliance with European Law. CA's exist in each European Member State and are nominated by each government to monitor and ensure compliance with the MDD (and upcoming MDR). The main role of the CA is to watch over the market in the member state for which it is designated. In particular, it must advise about the market launch of medical devices, advise about exporters and distributors performance, advise about any clinical studies being done with medical devices in that particular territory and advise about incidents that occur with medical devices. The CA designates a NB to ensure that conformity assessment procedures are completed as per relevant criteria.

A list of national competent authorities within the EU is provided below:

Austria
Federal Ministry of Health-Dept. Pharmaceuticals and Medical Devices
Radetzkystrasse 2, 1030 Wien/Vienna
E-mail: meddev@bmgf.gv.at
Federal Office for Safety in Health Care
Traisengasse 5, 1200 Vienna, Austria
Tel.: +43 50 555 36402
E-mail: medizinproduckte@basg.gv.at
Website: www.basg.gv.at

Belgium
Federal Agency for Nuclear Control
Health Protection Unit
Rue Ravenstein 36, 1000 Brussels
Tel.: +32 2 289 21 11
Fax: +32 2 289 21 12
E-mail: pointcontact@fanc.fgov.be
Website: www.afcn.fgov.be
Federal Agency for Medicines and Health Products
Health Products Division
Place Victor Horta 40/40, 1060 Brussels
Tel.: +32 2 524 80 00
Fax: +32 2 524 81 20
E-mail: meddev@afmps.be
Website: www.fagg-afmps.be

Bulgaria
Bulgarian Drug Agency
Department Control of Medicinal Products and Medical Devices
8, Damyan Gruev Str., 1303 Sofia, Bulgaria
Tel.: +359 2 890 34 83
Fax: +359 2 890 34 34
E-mail: bda@bda.bg
Website: www.bda.bg

Croatia
Ministry of Health, Health Protection Directorate
Ksaver 200a, 10000 Zagreb, Croatia
Tel.: +385 1 4607555
Fax: +385 1 4677076
Website: www.zdravlje.hr
Agency for Medicinal Products and Medical Devices
Ksaverska cesta 4, 10,000 Zagreb
Tel.: +385 1 4884 300
Fax: +381 1 4884 110
E-mail: halmed@halmed.hr
Website: www.halmed.hr

Cyprus
Ministry of Health
Cyprus Medical Devices Competent Authority
1 Prodromou & 17 Chilonos Street, 1448 Nicosia, Cyprus
Tel.: +357 22605572/735
Fax: +357 22468427
E-mail: cymda@mphs.moh.gov.cy
Website: www.moh.gov.cy/mphs
Czech Republic
State Institute for Drug Control
Šrobárova 48, 100 41 Praha 10.
Tel.: +420 272 185 111
E-mail: posta@sukl.cz

Denmark
Danish Health and Medicines Authority
Axel Heides Gade 1 DK-2300 Copenhagen S
Tel.: +45 72 22 74 00
Fax: +45 44 88 95 99
E-mail: med-udstyr@dkma.dk; dkma@dkma.dk
Website: www.medicinskudstyr.dk

Estonia
Health Board, Medical Devices Department
1a Põllu st., EE-Tartu 50303.
E-mail: MSO@terviseamet.ee
Website: www.terviseamet.ee

Finland
National Supervisory Authority for Welfare and Health (Valvira)
Mannerheimintie 103 b, P.O. Box 210, FI-00281 Helsinki, Finland
Tel.: +358 295 209 111
Website: www.valvira.fi

France
Ministry of the Economy and Finances
Direction générale de la concurrence, de la consommation et de la répression des frauds-Unit 5B
Télédoc 241-59, boulevard Vincent-Auriol F-75703 Paris Cedex 13
Tel.: +33 1 44 97 28 01
E-mail: Bureau-5B@dgccrf.finances.gouv.fr
Website: www.economie.gouv.fr/dgccrf
Agence nationale de sécurité du médicament et des produits de santé (ANSM)
143/147, boulevard Anatole France 93285 SAINT-DENIS CEDEX
Tel.: +33 1 55 87 36 86/+33 1 55 87 37 28
E-mail: dmdpt@ansm.sante.fr, dmtcos@ansm.sante.fr
Website: www.ansm.sante.fr
Direction Générale des douanes et droits indirects
Tel.: 0811 20 44 44/+33 1 72 40 78 50
E-mail: ids@douane.finances.gouv.fr
Website: www.douane.gouv.fr

Germany
Federal Institute for Drugs and Medical Devices
Kurt Georg Kiesinger Allee 3, D-53175 Bonn, Fax: +49 228 207 5300
E-mail: medizinprodukte@bfarm.de
Zentralstelle der Länder für Gesundheitsschutz bei Arzneimitteln und Medizinprodukten (ZLG)
E-mail: zlg@zlg.nrw.de
Federal Ministry of Health
Rochusstrasse 1, D-53123 Bonn
Fax:+49 228 941-4946
E-mail: medizinprodukte@bmg.bund.de

Greece
National Organization for Medicines
284 Mesogion Ave, GR – 15562 Holargos, Athens
E-mail: relation@eof.gr
Website: www.eof.gr

Hungary
Health Registration and Training Center (HRTC)
H-1085 Budapest, Horánszky Street 15
Tel.: +36 1 302 5060, Fax: +36 1 269 1255
E-mail: amd@enkk.hu
Website: www.enkk.hu

Iceland
Medicines Agency
Vínlandsleið 14, 113 Reykjavík
Tel.: 520 2100
Fax: 561 2170
E-mail: ima@ima.is
Website: www.imas.is

Ireland
Health Products Regulatory Authority (HPRA)
Kevin O'Malley House, Earlsfort Centre, Earlsfort Terrace, Dublin 2
Tel.: +353 1 676 4971
E-mail: info@hpra.ie
Website: www.hpra.ie

Italy
Ministry of Health, Department of Planning and Organization of the National Health Service-Directorate
General of Medical Devices, Pharmaceutical Services, and Safety in Healthcare
via Giorgio Ribotta 5, IT – 00144 Roma
E-mail: segr.dgfdm@sanita.it
Website: www.sanita.it

Latvia
Health Inspectorate
Klijanu Street 7, Riga, LV—1012, Latvia
Tel.: 371-67819671
Mob.: 371-26315189
Fax: 371-67819672
E-mail: vi@vi.gov.lv
Website: www.vi.gov.lv

Liechtenstein
Amt für Gesundheit
Äulestrasse 51, Postfach 684, 9490 Vaduz, Liechtenstein
E-mail: info.ag@llv.li
Website: www.ag.llv.li

Lithuania
State Health Care Accreditation Agency
Jeruzales St. 21 LT-08420 Vilnius Lithuania
Tel.: (85) 261 5177
Fax: (85) 212 7310
E-mail: vaspvt@vaspvt.gov.lt
Website: www.vaspvt.gov.lt

Luxembourg
Ministère de la Santé
Villa Louvigny-Allée Marconi L-2120 Luxembourg
Tel.: +352 24785500
E-mail: info@ms.public.lu

Malta
Malta Competition and Consumer Affairs Authority
Technical Regulations Division-Market Surveillance Directorate
Mizzi House, National Road, Blata l-Bajda, Hamrun, Malta HMR 9010
Tel.: +356 23952000
Fax: +356 21242406
E-mail: info@mccaa.org.mt
Website: www.mccaa.org.mt

Netherlands
The Health Care Inspectorate
Stadsplateau 1 3521 AZ Utrecht
Postbus 2680 3500 GR Utrecht
Tel.: +31 88 120 50 000
E-mail: meldpunt@igz.nl
Website: www.igz.nl

Norway
The Norwegian Medicines Agency
PO Box 240 Skøyen, 0213 Oslo, Norway
Tel.: +47 22 89 77 00
E-mail: meddev-no@noma.no
Website: www.legemiddelverket.no

Poland
Office for Registration of Medicinal Products, Medical Devices and Biocidal Products-Department of Medical
Devices Surveillance, Vigilance, and Clinical Trials
Al. Jerozolimskie 181C, 02-222 Warsaw, Poland
Tel.: +48 22 492 15 54
Fax: +48 22 492 11 09
E-mail: incydenty@urpl.gov.pl
Website: http://www.urpl.gov.pl/

Portugal
INFARMED–National Authority of Medicines and Health Products, IP
Parque de Saúde de Lisboa, Avenida do Brasil 53, 1749-004 Lisboa
Tel.: +351 217 987 100
Fax: +351 217 987 316
E-mail: infarmed@infarmed.pt
Website: www.infarmed.pt

Romania
National Agency for Medicines and Medical Devices
58, Sos. Nicolae Titulescu, Sector 1, Bucharest
Tel.: +4 021 2228652, +4 031 432 9120
Fax: +4 021 2228683, +4 031 432 9121
E-mail: mdevice@anm.ro
Website: www.anm.ro

Slovak Republic
National Institute for Drug Control
Kvetná 11 825 08 Bratislava 26 Slovak Republic
Tel.: +421-2-5070 11 11
Fax: +421-2-55 56 41 27
E-mail: sukl@sukl.sk
Website: www.sukl.sk

Slovenia
Agency for Medicinal Products and Medicinal Services
Ptujska ulica 21, 1000 Ljubljana, Slovenia
Tel.: +386 (0)8 2000 500
Fax: +386 (0)8 2000 630
E-mail: info@jazmp.si; meddev@jazmp.si
Website: www.jazmp.si

Spain
Agencia Española de Medicamentos y Productos Sanitarios
C/ Campezo 1, Edificio 8, ES – 28022 Madrid
E-mail: sgps@aemps.es
Website: http://www.aemps.gob.es/

Sweden
The Medical Products Agency
P.O. Box 26, SE-751 03 Uppsala
Tel.: +46 18 17 46 00
Fax: +46 18 50 31 15
E-mail: meddev.central@mpa.se
Website: www.lakemedelsverket.se
The Health and Social Care Inspectorate
P.O Box 45184, SE-104 30 Stockholm
Tel.: +46 10-788 50 00
E-mail: registrator@ivo.se
Website: www.ivo.se

Switzerland
Swissmedic, Swiss Agency for Therapeutic Products
Hallerstrasse 7, CH – 3012 Berne.
E-mail: medical.devices@swissmedic.ch
Website: https://www.swissmedic.ch

Turkey
Ministry of Health-Turkish Medicines and Medical Devices Agency
Vice Presidency of Inspectorate-Medical Devices Inspection Department
Sö ütözü Mahallesi 2176. Sokak No:5 PK 06520 Çankaya/ANKARA
Tel.: +90 312 218 35 34
Fax: +90 312 218 37 10
E-mail: tibbicihazpgd@titck.gov.tr
Website: http://www.titck.gov.tr

United Kingdom
Medicines and Healthcare Products Regulatory Agency
151 Buckingham Palace Road
London SW1W 9SZ
Tel.: +44 203 080 6000
Fax: +44 203 118 9803
E-mail: info@mhra.gsi.gov.uk

2.4.2. Notified Bodies (NB)

A notified body is defined as an independent certification organization that is "notified" by the competent authority (CA) of a European Member State. It plays an imperative role when it comes to supporting a manufacturers ability to place safe and compliant medical devices on the European market. The notified body is important for determining if a product or system meets the applicable requirements of the specific legislation/for CE marking and assists with carrying out conformity assessment procedures to demonstrate that the product complies with the requirements of the legislation. There are specific NBs dedicated to performing conformity assessments under Directive 93/42/EEC; however, it should also be noted that even fewer NBs are designated to perform conformity assessment under the Regulation (EU) 2017/745. Examples of these NBs are: BSI Assurance UK Ltd, BSI Group The Netherlands B.V., DEKRA Certification B.V, TUV SUD Rheinland and Intertek Medical Notified Body AB to name a few.

2.4.3. Authorized Representatives (AR)

The European AR (also known as the EU AR, EC REP, CE REP, EU REP or EAR) serves as a liaison between a manufacturer and the national CA and NB. The AR should be identified on the product labeling, outer packaging and the instructions for use (IFU) and will assist with device registrations. The AR will hold a copy of the manufacturer's technical file and/or CE declaration of conformity (DoC) and will make it available to the CA upon request. They will also assist with Incident and Field Safety Corrective Action reporting. In many cases, the AR will also provide manufacturers with regulatory news and relevant updates.

The Competent Authorities (CA), NB, and AR are all involved in the CE marking process which is interlinked between all three. The CA designates a NB to ensure that conformity assessment procedures are completed as per the relevant criteria. The authorized representative is designated by manufacturers and is legally responsible for compliance with the regulations. The AR acts as the first point of contact for the EU authorities. It is up to the manufacturer to ensure that their product complies with the essential requirements of the relevant EU legislation.

A general overview of the CE marking process is as in Figure 2.1.

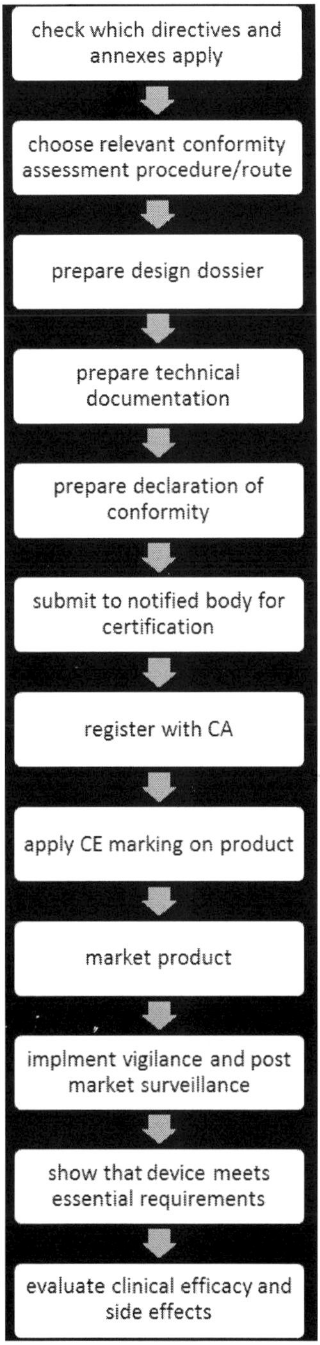

Figure 2.1: General overview of the CE marking process (easy-to-use CE marking infographics can be downloaded from https://scientistssanctuary.com/).

2.5. SIMPLIFIED PROCESS FOR CE MARKING OF MEDICAL DEVICES IN EUROPE

2.5.1. Step 1: Determine If the Product Concerned Is a Medical Device

A medical device is defined as a product or equipment that is intended for use within the health care industry to either diagnose, monitor or treat diseases or to act as a supportive aid for people with any form of illness or disability. There are over 500,000 different types of medical devices, and these include a vast array of devices such as x-ray machines, infusion pumps and cardiac pacemakers used by doctors, nurses, and surgeons as well as common devices that may be found every day in homes such as wound dressings, thermometers, contact lenses and wound irrigation solutions. According to ISO13485, medical devices range from simple low risk products such as urine bags and latex gloves to more complex, technologically advanced programmable devices such as pacemakers. Generally, these devices are grouped into four classes namely class I, class IIa, class IIb and class III; with class I being the lowest risk and Class III being the highest risk (Figure 2.2) or Class A, B, C, D with Class A being the lowest risk and Class D being the highest risk.

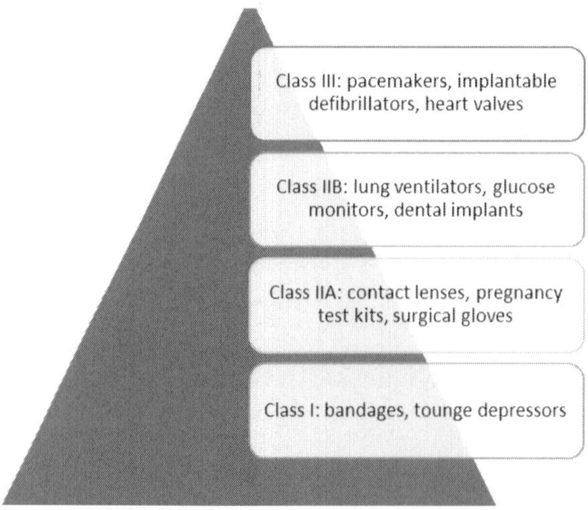

Figure 2.2: Overview of the different medical device classes.

The World Health Organization (WHO) defines a medical device as any instrument, apparatus, implement, machine, appliance, implant, reagent for *in vitro* use, software, material or other similarly related article intended by the manufacturer to be used, alone or in combination and for human beings in one or more specific medical purpose(s). These devices should not achieve its main intended function by pharmacological, immunological or metabolic means on or in the human body. It may, however, be assisted in its intended function by the above-mentioned methods.

2.5.1.1. General Uses for Medical Devices

- Diagnosis, prevention, monitoring, treatment or alleviation of disease.
- Diagnosis, monitoring, treatment, alleviation of, or compensation for an injury.
- Investigation, replacement, modification, or support of the anatomy and/or of a physiological process.
- Supporting or sustaining life.
- Control of conception.
- Disinfection of medical devices.
- Providing information by means of *in vitro* examination of specimens derived from the human body.
- Does not achieve its primary intended action by pharmacological, immunological or metabolic means, in or on the human body, but maybe assisted in its intended function by such means.

Source: Derived from WHO (https://www.who.int/medical_devices/full_definition/en/).

The mode of action of a medical device on the human body is what differentiates it from medicines, whose mode of action may often be metabolic, immunological or pharmacological. In fact, definitions must make clear distinctions on the difference between a medical device and medicine/drug whose primary intended use is obtained through chemical action or by being metabolized in the body. While many regulatory bodies tend to establish their own definitions around this topic, the global harmonization task force (GHTF) has proposed a harmonized definition for medical devices and *in-vitro* medical devices to ensure uniformity of understanding on a global scale (GHTF document SG1/N029R11).

2.5.1.2. Definition of a Medical Device as Defined by GHTF (GHTF/SG1/N071:2012)

"Medical device" means any instrument, apparatus, implement, machine, appliance, implant, *in vitro* reagent or calibrator, software, material or other similar or related article, intended by the manufacturer to be used, alone or in combination, for human beings for one or more of the specific purposes of:

- Diagnosis, prevention, monitoring, treatment or alleviation of disease;
- Diagnosis, monitoring, treatment, alleviation of or compensation for an injury;
- Investigation, replacement, modification, or support of the anatomy or of a physiological process;
- Supporting or sustaining life;
- Control of conception;
- Disinfection of medical devices;
- Providing information for medical purposes by means of *in vitro* examination of specimens derived from the human body. And which does not achieve its primary intended action in or on the human body by pharmacological, immunological or metabolic means, but which may be assisted in its function by such means.

The definition laid out by the GHTF covers a multitude of different medical devices, some of which are complex and reflects the latest advances in technological developments such as imaging equipment and implants; as well as those simple devices such as tongue depressors, thermometers, scales, and latex gloves. The GHTF document further defines those produces which may be considered to be medical devices in some jurisdictions while not classified in others. These include 1) disinfection substances, 2) aids for persons with disabilities, 3) devices incorporating animal and/or human tissues, and 4) devices for *in-vitro* fertilization or assisted reproduction technologies.

2.5.1.3. Definition of an In Vitro Medical Device as Defined by GHTF (GHTF/SG1/N071:2012)

'*In vitro* diagnostic (IVD) medical device' means a medical device, whether used alone or in combination, intended by the manufacturer for the *in-vitro* examination of specimens derived from the human body solely or principally to provide information for diagnostic, monitoring or compatibility purposes.

2.5.2. Step 2: Classification of the Medical Device

Once it has been determined that the product is a medical device, the next step is to determine what its classification is. A medical device may be classified as:

- Class I: Non-sterile, non-measuring;
- Class I: sterile, measuring;
- Class IIa;
- Class IIb;
- Class III.

Class III medical devices cover the highest risk products. It should also be remembered that the higher the classification, the greater the level of assessment required. Classification rules are found in Annex IX of the MDD 93/42/EEC Directive and Annex VIII of the MDR 2017/745. The classification annexes include definitions of the terminology being used in the classification rules.

Classification of a medical device will depend on a series of factors. This includes:

- How long the device is intended to be in continuous use;
- Whether or not the device is invasive or surgically invasive;
- Whether the device is implantable or active;
- Whether or not the device contains a substance.

The class of the medical device is determined by the manufacturer's assigned intended purpose for use. it is important to remember that the classification should not be assigned based on the classification of other similar products manufactured by the same manufacturer or different manufacturers. For example, two sutures manufacturers by different manufacturers may well have the same composition but different intended purposes for use and hence different classifications. For this reason, a manufacturer may successfully change the classification of a product by clearly defining on the label the intended purpose in such a way that the device falls into a higher or lower class.

2.5.3. Step 3: Quality Management System (QMS)

The next step for a manufacturer is to implement a QMS in accordance with the MDD/MDR. Most companies apply the ISO 13485 standard, which defines the QMS requirements for regulatory purposes. This standard is used for QMS s where an organization needs to demonstrate its ability to provide medical devices and related services that consistently meet customer and applicable regulatory requirements. It is important to note that compliance to ISO 13485 is not mandatory in all countries due to the presence of territory specific regulations. Nevertheless, certification in accordance with ISO 13485 often presumes compliance with certain annexes of the European MDD and now MDRs. In addition, depending on the class of the medical device in question, the requirements of the manufacturers QMS can be determined. For example:

- For Class 1 non-sterile, non-measuring devices (such as examination gloves)-manufacturers do not need to comply fully to a QMS, i.e., they do not need not be audited to a full QMS. This is a self-declared conformity.
- For Class 1 sterile, Class IIa, Class IIb, and Class III medical devices-manufacturers must operate under a QMS. Implementation of a QMS may take between 3–6 months and must be audited by a Notified Body which has been selected by the Manufacturer.

2.5.4. Step 4: Technical File

While the implementation of a QMS is in progress, the manufacturer may choose to prepare the technical file of the devices which they are manufacturing. The technical file provides detailed information demonstrating compliance with the MDD 93/42/EEC and now, MDR 2017/745. The technical file is required for all medical devices, including class I devices however the name and detail required for a Class III device will differ. For class III devices, it is usually known as a design dossier. The technical file and design dossiers are audited by independent companies, which are better known as NB. These NB have been authorized by the EU Member State Ministry of Health. The manufacturer will select its own Notified Body for review of its technical files for the audit of the implemented QMS.

2.5.5. Step 5: Appoint an Authorized Representative

Manufacturers selling into the EU are required to appoint an authorized representative who will represent them in Europe if their business is not located in Europe itself. An authorized representative is also known as the EC Rep. The EC Rep must be qualified to handle regulatory matters and will serve as the regulatory liaison between the Manufacturer and the European Ministry of Health of the respective European country, better known as the CA. Regardless of the classification of the medical device, manufacturers are required to appoint an EC Rep located in Europe. The ER Rep may also carry out other responsibilities as defined in the MDRs and as delegated by the Manufacturer. Tasks may include registration of medical devices, incident reporting, assistance with recalls and communication with the CA. The EC Rep will have full access to the Technical File or Design Dossier depending on the class of the device if a review is requested by the CA. The name and address of the EC Rep must appear on the labeling. It is also possible for manufacturers to consider appointing distributors to undertake the role of EC Reps. In this case, manufacturers may carefully choose the EC Rep base on capability and mandatory responsibilities. Distributors must be aware that the Technical Files, which is a proprietary document, must be made available when requested; however, this can present problems with multiple Distributors in many different countries.

2.5.6. Step 6: Audit of the QMS and Technical File/Design Dossier

The QMS implemented by a manufacturer must be audited and the technical file/design dossier (depending on the class of the product) must be reviewed by the notified body selected by the manufacturer. Once the review of the Technical File and audit of the QMS is successful, the Notified Body will issue the CE Marking Certification. For class I Non-Sterile, Non-Measuring Medical Devices, these devices are not issued with the Notified Body CE marking Certificate as conformity is based on self-certification. For all Class 1 Medical Devices, the device must be registered with the CA where the EC Rep is located. Most European countries do not require Class IIa, IIb, and III Medical Devices to be registered as it is already subjected to an annual audit by the Notified Body. However, Italy requires all Classes of Medical Devices to be registered.

2.5.7. Step 7: Declaration of Conformity (DoC)

The DoC is a legally binding document prepared by the manufacturer stating that the device is in compliance with the applicable directive. It is usually drawn up on a company letterhead and states that the company is in compliance with the relevant regulation and related ISO standard. The DoC shall also include the following information:

- Identification of the medical device;
- Name of the EC representative;
- Classification of the device;
- Process of compliance;
- Notified body details and CE marking certificate number.

The DoC should be signed by the company once compliance to EU Regulation can be shown.

2.5.8. Step 8: Labeling and Documentation

There are 31 European countries and 23 official languages, which can make labeling and documentation requirements a bit of a challenge for manufacturers. Although it is not explicitly addressed in the MDD 93/42/EEC, manufacturers must ensure that the labeling and documentation comply with the National Laws of the European country where the medical devices will be sold except if an exemption has been obtained from the CA. Labeling and documentation must be done in the official language of that state, and manufacturers must ensure that translation into the national language is in compliance with the national regulatory requirements.

2.5.9. Step 9: Monitor Safety and Efficacy

After CE marking has been obtained and devices are now placed on the EU market, manufacturers must continue to monitor the safety and efficacy of the device. This is usually done by appropriately handling incident report according to the EU requirements defined in the regulations. Manufacturers must establish an effective marketing surveillance system and review this periodically. Manufacturers are also expected to implement corrective actions as needed.

Chapter 3

Classification of a Medical Device

CONTENTS

Medical devices are generally divided into different classes based on their use and risk when in contact with a patient. This is done in an attempt to ensure the correct regulations and quality systems can be applied with the ultimate goal of ensuring safe and efficient use in or on patients. Some countries have their own separate classification system for different groups of medical devices, such as general devices, active implantable medical devices (AIMD), and *in vitro* diagnostic (IVD) medical devices, while other countries utilize one system for the classification of all medical devices. Regardless of the country or system used, all classification schemes tend to be risk-based. Having a suitable classification system in place ensures a more appropriate and graduated system of control which allows manufacturers themselves to classify their own devices rather than by one or two general committees, which in effect would take too long. This also enforces that manufacturers should be able to know as early as possible in which class their product falls. In the United States of America, for example, medical devices are classified as class I, II, and III, with class III devices representing the highest risk type device and requiring more overall control. Medical devices regulated with the USA are classified through an FDA classification database and are given a seven-digit number based on the category of the product. On the other hand, in the European Union (EU), general devices are classified as class I, class I sterile, class I measuring, class IIa, class IIb or class III; with class III devices representing the highest risk device. In addition, IVD devices have their own classification system while active implantable devices are not classified at all.

While having such a classification system in place is essential, it should also be noted that due to the vast number of variable involved as well as the changing nature of some of these variables, the mere implementation of a simple set of classification rules solely based on technical features is impossible. As a result, specifically in the EU, the European legislator has since established a classification concept which is based on the potential hazard related to the use and possible failure of various medical devices while at the same time taking into account technology used and any related standards, regulations, and/or health policy considerations. This has resulted in the establishment of a small set of criteria; namely the duration of contact with the body, degree of invasiveness, and local vs systemic effects, which can be combined in various ways to determine the class of the product. In addition, despite these various classification systems or existing rules which aim to classify the vast majority of medical devices, there are however, a small number of exceptions or confusing classification situations which may arise. Common examples include borderline devices between two classes, devices that cannot be classified by existing rules due to their unusual nature, or even situations in which classification may lead to the wrong conformity assessment level being chosen.

Due to the varying classification systems, the GHTF has developed a recommended classification system in which medical devices are divided into classes, namely A, B, C, and D, with class D representing the highest risk class. This global scheme is however a recommendation to regulatory authorities (Table 3.1).

Table 3.1: Proposed Classification System for Medical Devices (GHTF, 2006)

Class	**Risk Level**	**Device Examples**
A	Low risk	Surgical retractors/tongue depressors
B	Low-moderate risk	Hypodermic needles/suction equipment
C	Moderate-high risk	Lung ventilator/bone fixation plate
D	High risk	Heart valves/implantable defibrillator

Figure 3.1 shows an overall illustration of increasing levels of regulatory requirements as the device risk class increases. These regulatory controls may include: (1) operation of a quality system which is recommended for all medical devices; (2) technical data; (3) product testing using in-house or independent resources; (4) documentation of clinical evidence to support the manufacturer's claims; (5) the need for and frequency of independent external audit of the manufacturer's quality system; and (6) independent external review of the manufacturer's technical data.

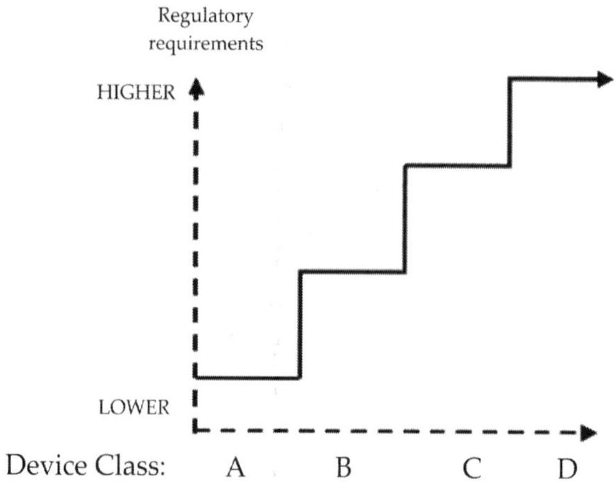

Figure 3.1: Illustration of various regulatory controls as per increasing risk class of medical devices.

3.1. FACTORS INFLUENCING DEVICE CLASSIFICATION SCHEMES

Several factors are known to influence medical device classification schemes. These range from the duration of contact which the device has with the body to the overall degree of invasiveness to whether the medical device is intended to have a biological or systemic effect on the patient. In some instances, the way in which a manufacturer defines the devices intended use may result in two or more classification rules being applicable to the device itself. In this instance, the highest level of classification will apply. If one medical device is intended to be used together with another medical device that may or may not be from the same manufacturer, then the classification rules should apply separately to each of the devices involved; for example, a general-purpose suture and a suture holder. In some instances, medical devices may be grouped and packaged together as per a manufacturer's requirements, and in this instance, classification of the medical device assemblage will depend on the manufacturer's purpose in packaging and marketing of such medical device combinations as when compared to individually packaged devices. For example, if the combination of devices together results in an overall product that is intended by the manufacturer to meet a specific purpose that is different from that of the individual medical devices that make up the combination, then the combination itself becomes a new medical device in its own right and should be classified according to the new intended use. However, on the other hand, if the medical device combination is for the convenience of the user but does not change the intended uses of the individual medical devices that make up that combination, then the classification allocated to that assemblage is at the level of the highest classified device within that group. On the other hand, if one or more medical devices within that assemblage of devices has yet to comply with all relevant regulatory requirements, then the combination should be classified based on the intended use of the device.

It is also important to remember that medical accessories that have been intended specifically to be used together with a 'parent' medical device such that it enables that medical device to achieve its intended purpose, should be subjected to all the GHTF guidance documents that apply to medical devices to ensure proper classification and application of relevant regulatory rules. These include, for example, GHTF documents around essential principles (EP) for Safety and Performance, post-market surveillance, medical device classification, and so on. in many instances, these medical device accessories may be classified as though they form actual medical devices in their own right. In other instances, while most software required for certain medical device operation may be incorporated into the medical device itself, some are not. If these standalone software falls within the scope of what is defined to be a medical device, then this software should be classified as follows:

- If the software drives or influences the use of a separate medical device, it should be classified according to the intended use of the combination;
- If the software is independent of any other medical device, it should be classified in its own right using the rules defined below in section x (taken from Section 8 of GHTF/SG1/N15:2006 Principles of Medical Device Classification, Study Group 1 as endorsed by The Global Harmonized Task Force, 2006);
- If the standalone software can be defined as being an active device.

3.2. INITIAL CLASSIFICATION RULES AS DEFINED BY THE GLOBAL HARMONIZATION TASK FORCE (GHTF)

The actual classification of each device depends on the intended use claims made by the manufacturer, and this indicates a need for manufacturers to thoroughly understand the purpose of each rule as defined below. For this reason, the actual classification of particular devices should be considered individually, taking into account its design and intended use (Table 3.2).

Table 3.2: Classification Rules as Defined by the GHTF

Rule	**Illustrative Examples of Devices That may Conform with a Rule**
Non-Invasive Devices	
➢ **Rule 1:** All non-invasive devices which come into contact with injured skin: • Are in Class A if they are intended to be used as a mechanical barrier, for compression or for absorption of exudates only, i.e., they heal by primary intent; • Are in Class B if they are intended to be used principally with wounds which have breached the dermis, including devices principally intended to manage the microenvironment of a wound. • Unless they are intended to be used principally with wounds which have breached the dermis and can only heal by secondary intent, in which case they are in Class C.	Devices covered by this rule are extremely claim sensitive. • Simple wound dressings; cotton wool. • Non-medicated impregnated gauze dressings. Devices used to treat wounds where the subcutaneous tissue is as least partially exposed, and the edges of the wound are not sufficiently close to be pulled together. To close the wound, new tissue must be formed within the wound prior to external closure. The device manufacturer claims that they promote healing through physical methods other than 'primary intent.' Examples include dressings for chronic ulcerated wounds; dressings for severe burns.
➢ **Rule 2:** All non-invasive devices intended for channeling or storing: • Body liquids or tissues. • Liquids or • Gases for the purpose of eventual infusion, administration or introduction into the body are in Class A unless they may be connected to an active medical device in Class B or a higher class, in which case they are Class B; unless they are intended for the use of: • Channeling blood, or • Storing or channeling other body liquids, or • For storing organs, parts of organs or body tissues, in which case they are Class B. unless they are blood bags, in which case they are Class C.	Such devices are 'indirectly invasive' in that they channel or store liquids that will eventually be delivered into the body (see comment for Rule 4). *Examples:* administration sets for gravity infusion; syringes without needles. *Examples:* syringes and administration set for infusion pumps; anesthesia breathing circuits. NOTE: "Connection" to an active device covers those circumstances where the safety and performance of the active device is influenced by the non-active device and vice versa. *Examples:* tubes used for blood transfusion, organ storage containers. *Example:* Blood bags that do not incorporate an anticoagulant. NOTE: in some jurisdictions, blood bags have a special rule that places them within a different risk class.

➢ **Rule 3:** All non-invasive devices intended for modifying the biological or chemical composition of: • Blood; • Other body liquids, or • Other liquids intended for infusion into the body are in Class C; unless the treatment consists of filtration, centrifuging or exchanges of gas or of heat, in which case they are in Class B.	Such devices are indirectly invasive in that they treat or modify substances that will eventually be delivered into the body (see note to comment for Rule 4). They are normally used in conjunction with an active device within the scope of either Rule 9 or 11. *Examples:* hemodialyzers; devices to remove white blood cells from whole blood. NOTE: for the purpose of this part of the rule, 'modification' does not include simple, mechanical filtration or centrifuging, which are covered below. *Examples:* devices to remove carbon dioxide; particulate filters in an extracorporeal circulation system.
➢ **Rule 4:** All other non-invasive devices are in Class A.	These devices either do not touch the patient or contact intact skin only. *Examples:* urine collection bottles; compression hosiery; non-invasive electrodes, hospital beds
Invasive Devices	
➢ **Rule 5:** All invasive devices with respect to body orifices (other than those which are surgically invasive) and which: • Are not intended for connection to an active medical device, or • Are intended for connection to a Class A medical device only. • Are in Class A if they are intended for transient use; • Are in Class B if they are intended for short-term use; unless they are intended for short-term use in the oral cavity as far as the pharynx, in an ear canal up to the eardrum or in a nasal cavity, in which case they are in Class A; • Are in Class C if they are intended for long-term use; unless they are intended for long-term use in the oral cavity as far as the pharynx, in an ear canal up to the eardrum, or in a nasal cavity and are not liable to be absorbed by the mucous membrane, in which case they are in Class B.	Section 4). Devices tend to be diagnostic and therapeutic instruments used in ENT, ophthalmology, dentistry, proctology, urology, and gynecology. Classification depends on the duration of use and the sensitivity (or vulnerability) of the orifice to such invasion. *Examples:* examination gloves; enema devices. *Examples:* urinary catheters, tracheal tubes. *Examples:* dentures intended to be removed by the patient; dressings for nose bleeds. *Example:* urethral stent; contact lenses for long-term continuous use (for this device, removal of the lens for cleaning or maintenance is considered as part of the continuous use). *Examples:* orthodontic wire, fixed dental prosthesis. *Examples*: tracheal tubes connected to a ventilator; suction catheters for stomach drainage; dental aspirator tips. NOTE: independent of the time for which they are invasive.

All invasive devices with respect to body orifices (other than those which are surgically invasive) that are intended to be connected to an active medical device in Class B or a higher class, are in Class B.	
➢ **Rule 6:** All surgically invasive devices intended for transient use are in Class B. • Unless they are reusable surgical instruments, in which case they are in Class A; or • Unless intended to supply energy in the form of ionizing radiation, in which case they are in Class C; or • Unless intended to have a biological effect or be wholly or mainly absorbed, in which case they are in Class C; or • Unless intended to administer medicinal products by means of a delivery system, if this is done in a manner that is potentially hazardous taking account of the mode of application, in which they are in Class C; or • Unless they are intended specifically for use in direct contact with the central nervous system, in which case they are in Class D; or • Unless intended specifically to diagnose, monitor or correct a defect of the heart or of the central circulatory system through direct contact with these parts of the body, in which case they are in Class D.	A majority of such devices fall into several major groups: those that create a conduit through the skin (e.g., syringe needles; lancets), surgical instruments (e.g., single-use scalpels; surgical staplers; single-use aortic punch); surgical gloves; and various types of catheter/sucker, etc. *Note*: a surgical instrument (other than those in Class D) is in Class A if reusable and in Class B if supplied sterile and intended for single use. Also, a surgical instrument connected to an active device is in a higher class than A. NOTE: if the device incorporates a medicinal substance in a secondary role, refer to Rule 13. *Examples:* Manually operated surgical drill bits and saws. *Example:* catheter incorporating/containing sealed radioisotopes. *Notes:* (a) the 'biological effect' referred to is an intended one rather than unintentional. The term 'absorption' refers to the degradation of a material within the body and the metabolic elimination of the resulting degradation products from the body. (b) This part of the rule does not apply to those substances that are excreted without modification from the body. *Example:* Insufflation gases for the abdominal cavity. *Example:* insulin pen for self-administration. NOTE: the term 'administration of medicines' implies storage and/or influencing the rate/volume of medicine delivered not just channeling. The term 'potentially hazardous manner' refers to the characteristics of the device and not the competence of the user. *Examples:* angioplasty balloon catheters and related guide wires; dedicated disposable cardiovascular surgical instruments.

➢ **Rule 7:** All surgically invasive devices intended for short-term use are in Class B. • Unless they are intended to administer medicinal products, in which case they are in Class C; or • Unless they are intended to undergo chemical change in the body (except if the devices are placed in the teeth), in which case they are in Class C; or • Unless they are intended to supply energy in the form of ionizing radiation, in which case they are in Class C; or • Unless they are intended to have a biological effect or to be wholly or mainly absorbed, in which case they are in Class D; or • Unless they are intended specifically for use in direct contact with the central nervous system, in which case they are in Class D; • Unless they are intended specifically to diagnose, monitor or correct a defect of the heart or of the central circulatory system through direct contact with these parts of the body, in which case they are in Class D.	Such devices are mostly used in the context of surgery or post-operative care, or are infusion devices, or are catheters of various types. *Examples:* infusion cannulae; temporary filling materials; non-absorbable skin closure devices; tissue stabilizers used in cardiac surgery. NOTE: includes devices that are used during cardiac surgery but do not monitor or correct a defect. *Note:* if the device incorporates a medicinal substance in a secondary role, refer to Rule 13. *Note:* the term 'administration of medicines' implies storage and/or influencing the rate/volume of medicine delivered not just channeling. *Example:* surgical adhesive. *Example:* brachytherapy device *Example:* absorbable suture; biological adhesive. *Note:* the 'biological effect' referred to is an intended one rather than unintentional. The term 'absorption' refers to the degradation of a material within the body and the metabolic elimination of the resulting degradation products from the body. *Example:* neurological catheter. *Examples:* cardiovascular catheters; temporary pacemaker leads; carotid artery shunts.
➢ **Rule 8:** All implantable devices, and long-term surgically invasive devices, are in Class C • Unless they are intended to be placed into the teeth, in which case they are in Class B; or • Unless they are intended to be used in direct contact with the heart, the central circulatory system or the central nervous system, in which case they are in Class D; or • Unless they are intended to be life-supporting or life-sustaining, in which case they are in Class D; or • Unless they are intended to be active implantable medical devices, in which case they are Class D; or	Most of the devices covered by this rule are implants used in the orthopedic, dental, ophthalmic, and cardiovascular fields. *Example:* maxilla-facial implants; prosthetic joint replacements; bone cement; non-absorbable internal sutures; posts to secure teeth to the mandibular bone (without a bioactive coating). *Note:* if the device incorporates a medicinal substance in a secondary role, refer to Rule 13. *Examples:* bridges, crowns; dental filling materials. *Examples:* prosthetic heart valves; spinal and vascular stents. *Example:* pacemakers, their electrodes and their leads; implantable defibrillators.

• Unless they are intended to have a biological effect or to be wholly or mainly absorbed, in which case they are in Class D; or • Unless they are intended to administer medicinal products, in which case they are in Class D; or • Unless they are intended to undergo chemical change in the body (except if the devices are placed in the teeth), in which case they are in Class D; or • Unless they are breast implants, in which case they are in Class D.	*Example:* implants claimed to be bioactive. *Note:* hydroxy-apatite is considered as having biological effect only if so claimed and demonstrated by the manufacturer. *Example:* rechargeable non-active drug delivery system. *Note:* bone cement is not within the scope of the term 'chemical change in the body' since any change takes place in the short rather than long term.
Active Devices	
➢ **Rule 9(i):** All active therapeutic devices intended to administer, or exchange energy are in Class B • Unless their characteristics are such that they may administer or exchange energy to or from the human body in a potentially hazardous way, including ionizing radiation, taking account of the nature, the density, and site of application of the energy, in which case they are in Class C.	Such devices are mostly electrically powered equipment used in surgery; devices for specialized treatment and some stimulators. *Examples:* muscle stimulators; TENS devices; powered dental handpieces; hearing aids; neonatal phototherapy equipment; ultrasound equipment for physiotherapy. *Examples:* lung ventilators; baby incubators; electrosurgical generators; external pacemakers and defibrillators; surgical lasers; lithotripters; therapeutic X-ray and other sources of ionizing radiation. NOTE: the term 'potentially hazardous' refers to the type of technology involved and the intended application.
➢ **Rule 9(ii):** All active devices intended to control or monitor the performance of active therapeutic devices in Class C, or intended directly to influence the performance of such devices, are in Class C.	*Examples:* external feedback systems for active therapeutic devices.

➢ **Rule 10(i):** Active devices intended for diagnosis are in Class B: • If they are intended to supply energy which will be absorbed by the human body (except for devices used solely to illuminate the patient's body, with light in the visible or near infra-red spectrum, in which case they are Class A), or • If they are intended to image *in vivo* distribution of radiopharmaceuticals, or • If they are intended to allow direct diagnosis or monitoring of vital physiological processes; unless they are specifically intended for: • monitoring of vital physiological parameters, where the nature of variations is such that it could result in immediate danger to the patient, for instance variations in cardiac performance, respiration, activity of central nervous system, or • Diagnosing in clinical situations where the patient is in immediate danger, in which case they are in Class C	Such devices include equipment for ultrasonic diagnosis/imaging, capture of physiological signals, interventional radiology and diagnostic radiology. *Examples:* magnetic resonance equipment; diagnostic ultrasound in non-critical applications; evoked response stimulators. *Examples:* gamma/nuclear cameras *Example:* electronic thermometers, stethoscopes, and blood pressure monitors; electrocardiographs. *Example:* monitors/alarms for intensive care; biological sensors; oxygen saturation monitors; apnoea monitors. *Example:* ultrasound equipment for use in interventional cardiac procedures.
➢ **Rule 10(ii):** Active devices intended to emit ionizing radiation and intended for diagnostic and/or interventional radiology, including devices which control or monitor such devices, or those which directly influence their performance, are in Class C.	*Example:* these include devices for the control, monitoring or influencing of the emission of ionizing radiation.
➢ **Rule 11:** All active devices intended to administer and/or remove medicinal products, body-liquids or other substances to or from the body are in Class B • Unless this is done in a manner that is potentially hazardous, taking account of the nature of the substances involved, of the part of the body concerned and of the mode and route of administration, in which case they are in Class C.	Such devices are mostly drug delivery systems or anesthesia equipment. Examples of Class B devices: suction equipment; feeding pumps; jet injectors for vaccination; nebulizer to be used on conscious and spontaneously breathing patients where failure to deliver the appropriate dosage characteristics is not potentially hazardous. *Examples:* infusion pumps; anesthesia equipment; dialysis equipment; hyperbaric chambers; nebulizer where the failure to deliver the appropriate dosage characteristics could be hazardous.

➢ **Rule 12:** All other active devices are in Class A	*Examples:* examination lamps; surgical microscopes; powered hospital beds and wheelchairs; powered equipment for the recording, processing, viewing of diagnostic images; dental curing lights.
Additional Rules	
➢ **Rule 13:** All devices incorporating, as an integral part, a substance which, if used separately, can be considered to be a medicinal product, and which is liable to act on the human body with action ancillary to that of the devices, are in Class D.	These medical devices incorporate medicinal substances in an ancillary role. *Examples:* antibiotic bone cements; heparin-coated catheters; wound dressings incorporating antimicrobial agents to provide ancillary action on the wound; blood bags incorporating an anticoagulant. NOTE: Such medical devices may be subject to additional conformity assessment procedures according to the regional or national requirements of medicinal product Regulatory Authorities.
➢ **Rule 14:** All devices manufactured from or incorporating animal or human cells/tissues/derivatives thereof, whether viable or non-viable, are Class D; • Unless such devices are manufactured from or incorporate non-viable animal tissues or their derivatives that come in contact with intact skin only, where they are in Class A.	*Note:* In some jurisdictions such products: are considered to be outside the scope of the medical device definition; • May be subject to different controls • It is likely the regulations controlling these devices will be the subject of future harmonization efforts. *Examples:* porcine heart valves; catgut sutures *Examples:* leather components of orthopedic appliances.
➢ **Rule 15:** All devices intended specifically to be used for sterilizing medical devices, or disinfecting as the endpoint of processing, are in Class C. • Unless they are intended for disinfecting medical devices prior to endpoint sterilization or higher-level disinfection, in which case they are in Class B; or • Unless they are intended specifically to be used for disinfecting, cleaning, rinsing or, when appropriate, hydrating contact lenses, in which case they are in Class C.	*Examples:* devices for disinfecting or sterilizing endoscopes; disinfectants intended to be used with medical devices. NOTE: This rule does not apply to products that are intended to clean medical devices by means of physical action, e.g., washing machines. *Example:* washer-disinfectors In some jurisdiction's solutions for use with contact lenses: are considered to be outside the scope of the medical device's definition; may be subject to different controls.
➢ **Rule 16:** All devices used for contraception or the prevention of the transmission of sexually transmitted diseases are in Class C • Unless they are implantable or long-term invasive devices, in which case they are in Class D.	*Examples:* condoms, contraceptive diaphragms *Example:* intrauterine contraceptive device

3.3. EU GUIDELINES ON MEDICAL DEVICE CLASSIFICATION

Detailed explanations of the European classification systems can be found in MEDDEV 2.4/1. General requirements of this classification scheme are such that all devices must:

- Meet the essential requirements irrespective of the class of the device;
- Be subject to the reporting requirements under the medical device vigilance system;
- Be CE marked (except for custom-made devices and devices that are intended for clinical investigation).

This medical device classification systems are needed to channel medical devices into the proper conformity assessment route (Table 3.3) as subjecting all existing medical devices to rigorous conformity assessment procedures would just not be feasible economically.

Table 3.3: Conformity Assessment Procedures vs. Classes

Conformity Assessment Procedures	**Classes**					
Annexes	**I**	**I Sterile**	**I Measuring**	**IIA**	**IIB**	**III**
II (+ sect. 4)						√
II (-sect 4)				√	√	
III					√	√
IV		√	√	√	√	√
V		√	√	√	√	√
VI		√	√	√	√	
VII	√	√	√	√		

When carrying out a classification as per MEDDEV 2.4/1, the manufacturer should first decide if the product is a medical device as defined by Directive 93/42 or if it is an accessory to a medical device. Determining the exact grouping of the device will aid a manufacturer in knowing which directive to apply. Active implantable devices and devices for *in vitro* diagnosis are covered by separate directives which in actual fact do not apply the classification rules from guidelines supplied by MEDDEV 2.4/1, thereby further indicating that different groups of devices will utilize different rules. Classification rules as defined by MEDDEV 2.4/1 are based on terms related to 1) the duration of contact with the patient, 2) degree of invasiveness, and 3) the part of the body affected by the use of the device itself. One of the first steps to applying the classification rules is understanding the basic definitions used in the guideline, all of which are defined in Section I of Annex IX of the Directive.

3.3.1. Duration

The duration of contact with the patient, also known as *time* can be classified into transient, short term or long term. Transient is defined as that which is normally intended for continuous use for *less than 60 minutes* while short term is defined as that which is normally intended for continuous use for *not more than 30 days* and long term is that which is normally intended for continuous use *for more than 30 days*.

3.3.2. Continuous Use

In addition, understanding the concept of continuous use is important when the intended use of the device is taken into consideration. The abovementioned concept of duration (transient, short term and long term) are all defined in terms of continuous use which is better defined as an uninterrupted actual use for the intended purpose of the device in question. For example, a scalpel which may be used on a patient for cutting of tissue will not normally last for more than a few seconds at a time and can therefore be defined as a continuous transient use device. On the other hand, a device that is used and replaced immediately either by the same or identical device is considered as discontinued and shall be defined as an extension of the continuous use of the device. A common example would be a suture needle which is used for stitching wounds.

3.3.3. Invasiveness

Invasive devices are defined as a device which either, completely or partly, penetrates inside the body. This usually occurs either through a body orifice (better known as any natural opening in the body) or through the surface of the body. An invasive device which penetrates inside the body through the surface of the body, with the aid or in the context of a surgical operation is better known as a surgically invasive device. Thus, a surgically invasive device always implies that it enters through an artificially created opening, which could be a large opening, such as a surgical incision, or as small as a needle hole created by a suture needle.

3.3.4. Implantable Device

An implantable device is defined as any device which is intended for complete introduction into the human body or which will be replaced by an epithelial surface or the surface of the eye; either by surgical intervention and which is intended to remain in place after the procedure. In addition, any device intended to be partially introduced into the human body through surgical intervention and intended to remain in place after the procedure for at least 30 days is also considered an implantable device. When defining an implantable device, the idea behind the term 'procedure' must be taken into consideration and as such, an implantable device must remain in the patient after the procedure is complete. As such, a procedure must include the surgical activity during which the implantable device is placed in the body and the immediate post-operative procedures associate with the initial procedure itself. That being said, the term procedure does not include the removal of an implant, which in effect must be considered to be another procedure. As a result, the placing of an implant and the removal of the implant would be become two different surgical procedures. Furthermore, some partially implanted devices are also classified to be implants, but clear distinction is needed.

3.3.5. Active Devices

Active medical devices are defined as any medical device in which their operation depends on a source of electrical energy or any power source other than that directly generated by the human body or gravity and which acts by converting this energy. Those medical devices that aim to simply transmit energy, substances or other elements between an active medical device and the patient themselves, without any significant change, are not considered to be active medical devices. In order for this definition to be conveyed properly, the concept of "act by converting energy" and "significant changes" must be clearly explained and understood. To "act by converting energy" includes the conversion of energy in the device and/or conversion at the interface between the device itself and the tissues or within the tissues. The concept of "significant changes" on the other hand, includes changes in the nature, level, and density of energy. For example, an electrode would not be defined as an active medical device as per this definition as long as the energy input is intended to be the same as the energy output. On the other hand, electrodes used in electrosurgery for cutting tissues can be defined as active medical devices as per the above definition as their operation depends on energy provided by a generator and their action is achieved by conversion of the energy at the interface between the device and the tissue itself or within the tissue. However, those electrodes that are intended for E.C.G or E.E.G's are to normally define as active devices because they do not normally act by conversion of energy. Manufacturers should however, consult the "Guidelines relating to the application of the Council Directive 90/385/EEC on AIMD (Med.Dev. 2.1/2)" should they require more information or better understanding on the issue.

3.3.5.1. Devices with a Measuring Function

Devices are considered to have a measuring function of the measure a parameter such as volume or a physiological parameter and display or indicate its value in a defined unit of measurement. Common examples include urine bags or thermometers.

Certain criteria have been laid out as per MEDDEV 2.1/5 (1998) which a Guidance document relating to the application of medical devices with a measuring function. This guidance document states criteria which if fulfilled together, indicate that a device does, in fact, have a measuring function. These criteria are defined below:

1. The device is intended by the manufacturer to measure:
   - Quantitatively a physiological or anatomical parameter.
   - A quantity or a quantifiable characteristic of energy or of substances delivered to or removed from the human body.
2. The result of the overall measurement:
   - Is displayed in legal units or other acceptable units within the meaning of Directive 80/181/ECC[1].
   - Is compared to at least one point of reference indicated in legal units or other acceptable units in compliance with the pre-mentioned directive.
3. The intended purpose implies accuracy (claimed either explicitly or implicitly) where any form of non-compliance with the implied accuracy could result in a significant adverse effect on the patient's health and safety.

Note 1: The phrase "claimed implicitly" covers cases where the user, on the basis of the designation of the device or of its accompanying documents, or on the basis of the common use is entitled to expect accuracy where the accuracy of the measurement has an impact on the diagnosis or therapy of the patient (Table 3.4).

Table 3.4: Common Examples of Devices with a Measuring Function

Criteria	Example
Devices used for measuring body temperature	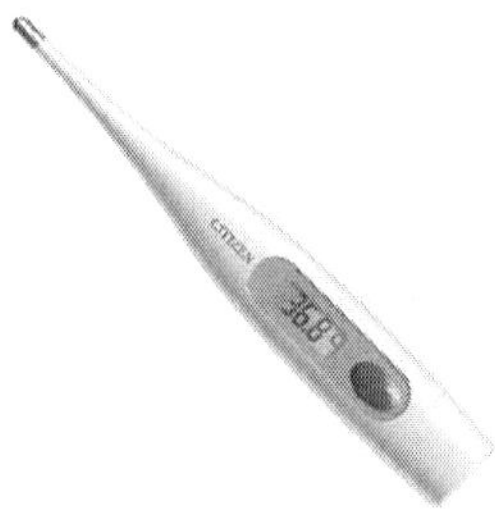
Pacifiers which include a temperature display including those with only a change of color where criteria b is met	

Devices for indicating that body temperature is above or below a specified value	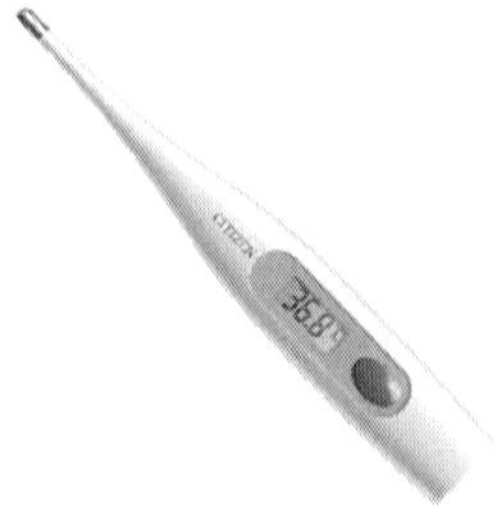
Non-active non-invasive devices for measuring blood pressure	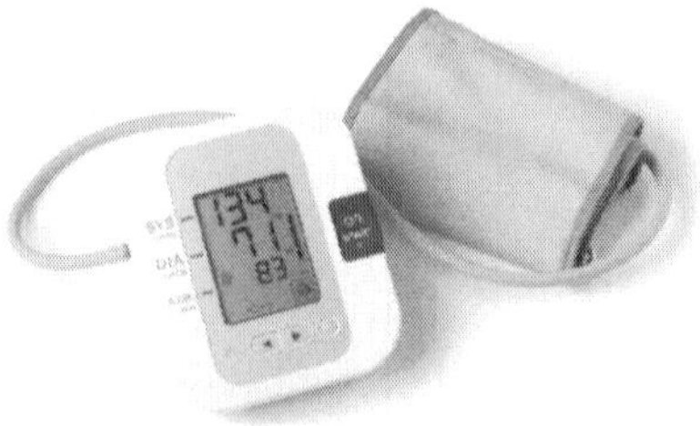
Non-active devices for measuring intra-ocular pressure	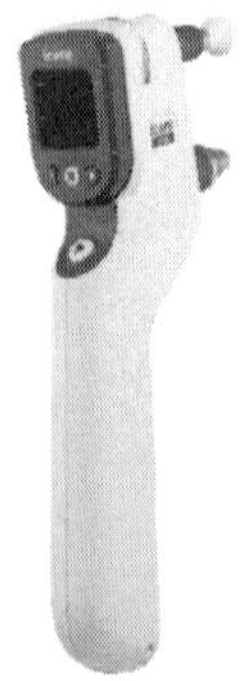
Devices for measuring volume or pressure or flow of liquid or gases delivered to or removed from the human body. This includes any container with a graduation scale or with a single point graduation where criteria c is met.	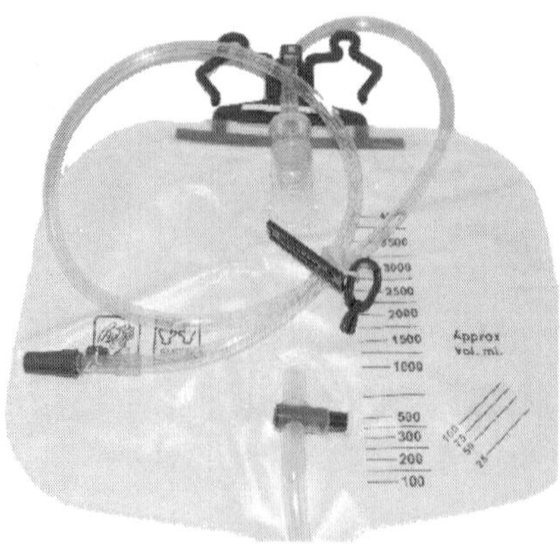

3.3.5.2. Devices without a Measuring Function (Table 3.5)

Table 3.5: Common Examples of Devices Without a Measuring Function

Criteria	Example
Devices used for the delivery of liquid to the human body (for example, medicine spoons, cups, droppers, without graduation or scale or display of measuring unit).	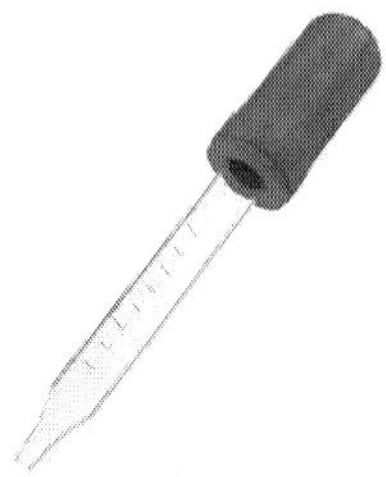
Devices used for displaying trends of physiological parameters (for example, urine bags without graduation or scale, calipers for obesity).	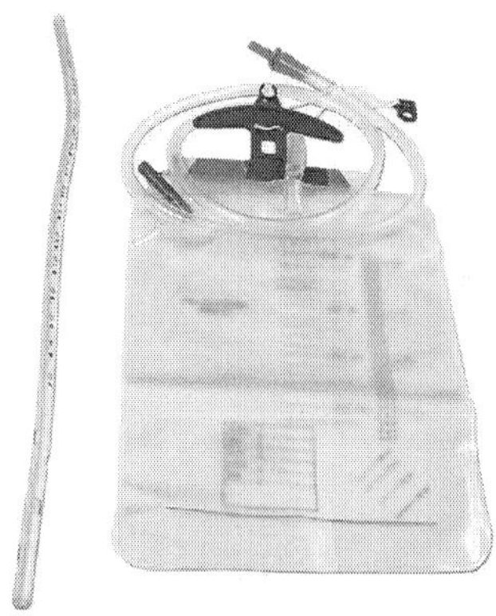
Eye-test charts	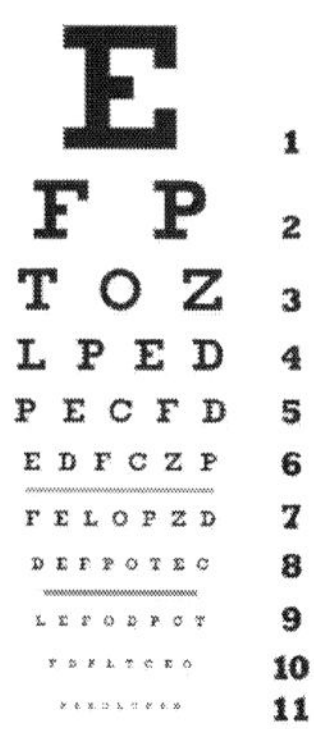

3.4. UNDERSTANDING HOW TO APPLY THE RULES OF MEDICAL DEVICE CLASSIFICATION

When applying the classification rules, the following should be considered:

- The intended purpose of the device as opposed to technical characteristics as this will determine the class of the device itself (unless these have a direct influence on the intended purpose).
- The intended use must not be confused with common accidental uses of the device.
- The intended purpose assigned by the actual manufacturer to the specific device should determine the class of the device rather than allowing for classification by inferences such as the class of other similar products determining the class of a specific device.
- Classification of device components in relation to the system as a whole or individually should be taken into consideration. For example, combination devices with parts that have different functional purposes may be analyzed separately with respect to each of the parts; such as a drainage device that has an invasive tube and a non-invasive collection device. Each of these components may be classified separately.
- Accessories must be classified separately from their parent devices.
- If a device can be classified by several rules, then the highest possible class should apply.
- If a device is not intended to be used solely in a specific part of the body, then it must be considered and classified based on the most critical specified use. As such classification of the device will have to be determined based on claims contained in the information supplied with the medical device itself. For this reason, manufacturers must be extremely specific with regards to the overall intended use while also providing as a minimum requirement, either appropriate positive or negative indications for use.
- Multi-application equipment (such as laser printers or identification cameras) which may be used in combination with medical devices are not actual medical devices unless their manufacturer places them on the market with specific intended purpose as a medical device or for use with medical devices only.

- Software that is regarded as driving or influencing the use of a medical device may fall in the same class of the device which it influences. If, however, such standalone software is not regarded as driving or influencing the use of a medical device, then it should be classified separately and in its own right.

3.5. HOW TO APPLY THE EU MEDICAL DEVICE CLASSIFICATION RULES?

In order to appropriately apply the EU medical device rules for classification, manufacturers must take into account all the rules laid out in the guidelines to ensure their devices are properly classified. In addition, all device characteristics or combination of characteristics in relation to its intended purpose must be taken into consideration. One such classification example is defined below:

A wound drainage device produced by Manufacturer X needs to be classified:

A simple wound drainage device has three components that must be taken into consideration, namely the cannula, the tubing, and the collector unit. If the particular device in question is to be sold without a cannula, then the classification of the cannula does not need to be taken into consideration. The manufacturer has also stated that the device will be used under short-term duration, i.e., uninterrupted intended use will be for more than 60 minutes and less than 30 days. It also assumed that any collected liquids are not intended to be re-infused into the body nor reprocessed for eventual re-infusion and that the device is not intended to be connected to a powered suction system (Table 3.6).

Table 3.6: Intended Uses of Devices and Their Risk Classification

Intended Uses	Rule	Class
Surgically invasive cannula to reach a wound site in the pleural cavity to drain the cavity	7	IIA
Non-invasive tubing to evacuate body-liquids towards the collector	1	I
Non-invasive collector to receive the body liquids	1	I

From the above defined specifications by the manufacturer, it is clear that they would have a choice of applying a Class II A risk level to the whole device. Alternatively, they would have to carry out a separate conformity assessment procedure for the cannula and another conformity assessment procedure for the tubing and another for the collector.

As defined by MEDDEV 2.4/1, different rules will apply to different groups of devices (Table 3.7). As a rule of thumb, however, it should be remembered that the characteristics or combination of characteristics in accordance with its intended purpose of the device that determines the highest class will determine the overall class for the device (Figures 3.2–3.6).

Table 3.7: Different Rules That Apply for Risk Classification of Different Devices

Type of Device	Applicable Rules as per MEDDEV 2.4/1 Classification Rules
Non-invasive	1,2,3,4
Invasive	5,6,7,8
Active	9,10,11,12
Special	13,14,15,16,17,18

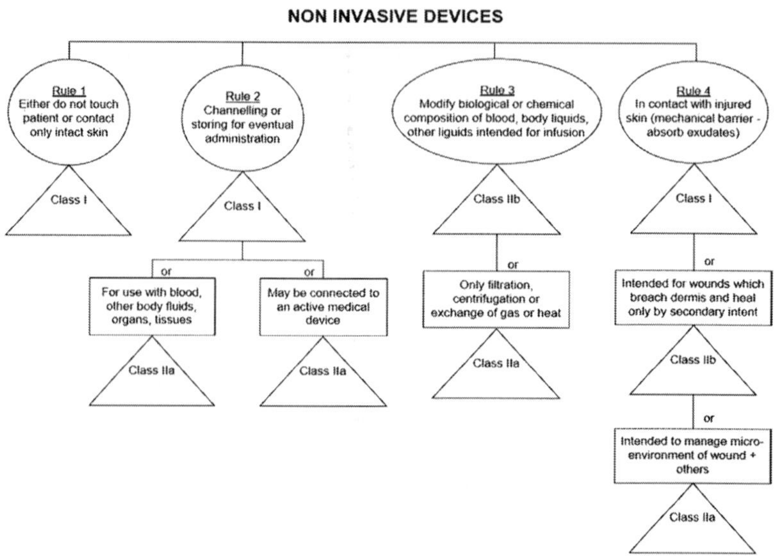

Figure 3.2: Flow diagram showing risk classification routes based on Rules 1–4 from MEDDEV2.4/1.

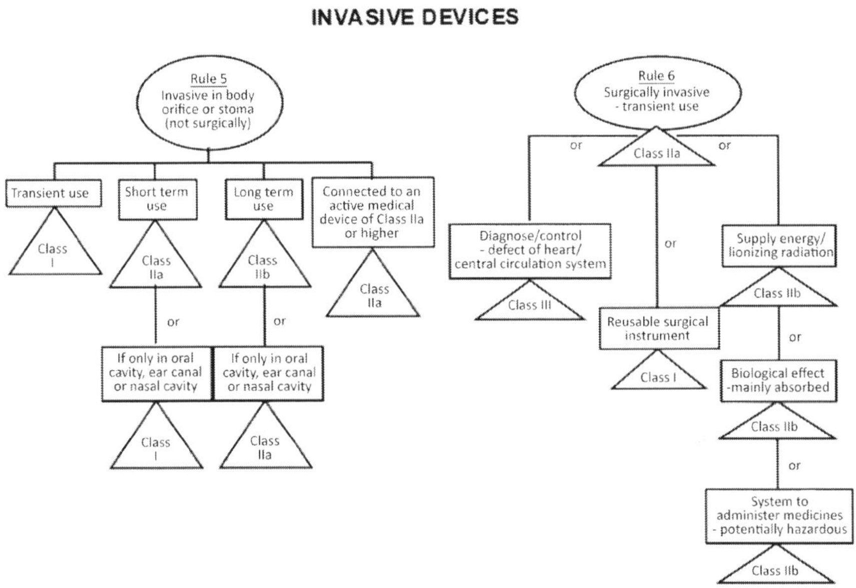

Figure 3.3: Flow diagram showing risk classification routes based on Rules 5–6 from MEDDEV2.4/1.

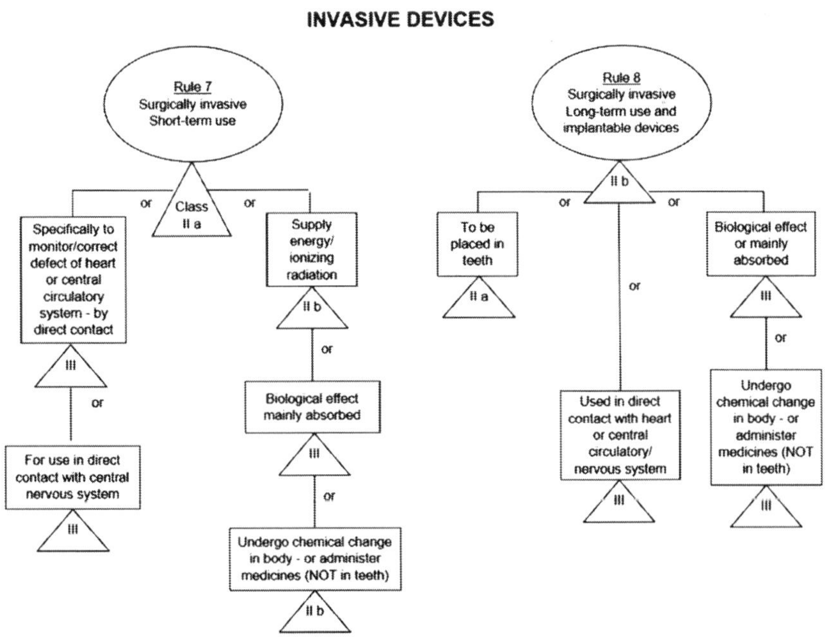

Figure 3.4: Flow diagram showing risk classification routes based on Rules 7–8 from MEDDEV2.4/1.

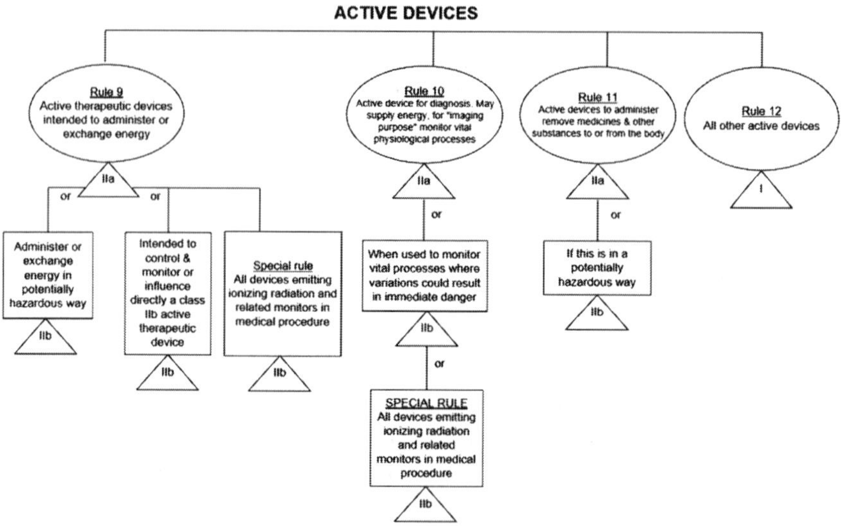

Figure 3.5: Flow diagram showing risk classification routes based on Rules 9–12 from MEDDEV2.4/1.

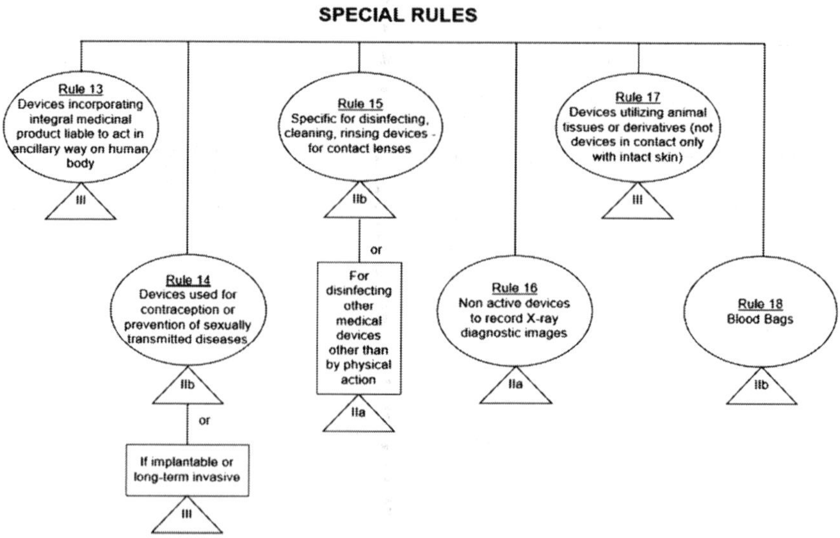

Figure 3.6: Flow diagram showing risk classification routes based on Rules 13–17 from MEDDEV2.4/1.

3.6. UNDERSTANDING THE RULES GOVERNING GUIDELINES FOR THE CLASSIFICATION OF MEDICAL DEVICES ACCORDING TO MEDDEV 2.4/1

The below sub-chapter covers a general explanation of the rules and practical issues governing the EU guidelines for medical device classification as laid out in MEDDEV 2.4/1.

3.6.1. Rule 1: Devices That Either Do Not Touch the Patient or Contact Intact Skin Only

General explanation of the rule: Rule 1 forms a fallback rule that is applicable to all medical devices that are not generally covered by a more specific rule later. It applies in general to all devices that meet only intact skin, and which does not touch the patient (Table 3.8).

Table 3.8: Classification Rules as Per Rule 1 of MEDDEV 2.4/1

Rule 1	Examples
All non-invasive devices are in Class I, unless one of the rules set out hereinafter applies.	Body liquid collection devices intended to be used in such a way that a return flow is unlikely (e.g., to collect body wastes such as urine collection bottles, ostomy pouches, incontinence pads or collectors used with wound drainage devices). They may be connected to the patient by means of catheters and tubing. Devices used to immobilize body parts and/or to apply force or compression on them (e.g., non-sterile dressings used to aid the healing of a sprain, plaster of Paris, cervical collars, gravity traction devices, compression hosiery). Devices intended in general for external patient support (e.g., hospital beds, patient hoists, walking aids, wheelchairs, stretchers, dental patient chairs). Corrective glasses, frames, stethoscopes for diagnosis, eye occlusion plasters, incision drapes, conductive gels, non-invasive electrodes (electrodes for EEG or ECG), image intensifying screens. Permanent magnets for removal of ocular debris

- **Understanding the Practical Issues Around Rule 1 Classification:** Some non-invasive devices are indirectly in contact with the human body and can influence internal physiological processes by storing, channeling or treating blood, other body liquids or liquids which are returned or infused into the body or by generating energy that is delivered to the body. These must be excluded when applying this rule and should be handled by another rule because of the hazards inherent in such direct influence on the body.

3.6.2. Rule 2: Channeling or Storing for Eventual Administration

General explanation of the rule: these types of devices must be considered separately from the non-contact devices of rule 1 because they may be indirectly invasive. The channel or store substances that will be eventually delivered into the body. Typically, these devices after used in transfusion, infusion, extracorporeal circulation, delivery of anesthetic gases and oxygen. In some cases, devices covered under this rule are very simple devices that are gravity activated (Table 3.9).

Table 3.9: Classification Rules as Per Rule 2 of MEDDEV 2.4/1

Rule 2	Examples
All non-invasive devices intended for channeling or storing blood, body-liquids or tissues, liquids or gases for the purpose of eventual infusion, administration or introduction into the body are in Class IIa: If they may be connected[1] to an active medical device in Class IIa or a higher class.	Devices intended to be used as channels in active drug delivery systems, e.g., tubing intended for use with an infusion pump. Devices used for channeling, e.g., antistatic tubing for anesthesia, anesthesia breathing circuits and pressure indicator, pressure limiting devices. Syringes for infusion pumps.
If they are intended for use for storing or channeling blood or other body liquids or for storing organs, parts of organs or body tissues (are in Class II a).	Devices intended to channel blood (e.g., in transfusion, extracorporeal circulation). Devices intended for temporary storage and transport of organs for transplantation. Devices intended for long term storage of biological substances and tissues such as corneas, sperm, human embryos, etc.

in all other cases they are in Class I	Devices that provide a simple channeling function, with gravity providing the force to transport the liquid, e.g., administration sets for infusion. Devices intended to be used for a temporary containment or storage function such as cups and spoons specifically intended for administering medicines[2]. Syringes without needle

- **Understanding the Practical Issues Around Rule 2 Classification:** If a device can be used for a purpose that would result it to be connected to an active device, then such a device will automatically fall in Class II A unless the manufacturer clearly states that it should not be connected to an active device of Class II-A or higher. It should also be noted that within Rule 2, blood bags are covered as an exception under a separate rule (Rule 18).

Special concepts are also noted in Table 3.9 for Rule 2 and are explained as follows:

- *Note 1:* All non-invasive devices intended for channeling or storing blood, body-liquids or tissues, liquids or gases for the purpose of eventual infusion, administration or introduction into the body are in Class IIa: *If they may be connected[1] to an active medical device in Class IIa or a higher class;*

Such connection is deemed to exist between a non-active device and an active device where a non-active device forms a link in the transfer of the substance between the patient and the active device. Furthermore, the safety and performance of one of the devices is influenced by the other device.

An example as explained in the guideline, would be tubing in an extracorporeal circulation system which is downstream from a blood pump and in the same blood flow circuit but not directly in contact with the pump.

- *Note 2:* "Devices intended to be used for a temporary containment or storage function such as cups and spoons specifically intended for administering medicines."

This statement refers to solutions intended for the preservation of organs during storage and transport and not medical devices.

3.6.3. Rule 3: Devices That Modify Biological or Chemical Composition of Blood, Body Liquids or Other Liquids

General explanation of the rule: these types of devices must be considered separately from non-contact devices of rule 1 because they are indirectly invasive and treat or modify substances that will eventually be delivered into the body. Rule 3 covers mostly the more sophisticated elements of extracorporeal circulation sets, dialysis systems and autotransfusion systems as well as devices for extracorporeal treatment of body fluids which may not be reintroduced immediately into the body, including where the patient is not in a closed loop with the device (Table 3.10).

Table 3.10: Classification Rules as Per Rule 3 of MEDDEV 2.4/1

Rule 3	Examples
All non-invasive devices intended for modifying the biological or chemical composition of blood, other body liquids or other liquids intended for infusion into the body are in Class IIb	Devices intended to remove undesirable substances out of the blood by exchange of solutes such as hemodialyzers. Devices intended to separate cells by physical means, e.g., gradient medium for sperm separation. Hemodialysis concentrates
unless the treatment consists of filtration, centrifugation or exchange of gas or heat, in which case they are in Class IIa	
unless the treatment consists of filtration, centrifugation or exchange of gas or heat, in which case they are in Class IIa	Particulate filtration of blood in an extracorporeal circulation system. These are used to remove particles and emboli from the blood. Centrifugation of blood to prepare it for transfusion or autotransfusion. Removal of carbon dioxide from the blood and/or adding oxygen. Warming or cooling the blood in an extracorporeal circulation system

- **Understanding the Practical Issues of Rule 3 Classification:** These devices are normally used in conjunction with an active medical device covered under rule 9 or rule 11. Filtration and centrifugation should be understood in the context of this rule as exclusively mechanical methods.

3.6.4. Rule 4: Devices in Contact with Injured Skin

General explanation of the rule: Rule 4 is intended to cover mainly wound dressings regardless/independent of the wound. Traditional types of products such as those commonly used as mechanical barriers, are often well understood and do not result in any great hazard. In addition, there have been rapid technological developments in this area, with the emergence of new types of wound dressings for which non-traditional claims are made. Examples of non-traditional claims include management of the micro-environment of a wound to enhance its natural healing mechanisms. Other ambitious claims may also be included, and these involve the mechanism of healing by secondary intent. This rule also includes some devices used on breached dermis which may even have a life-sustaining or lifesaving purpose, such as when there is full-thickness destruction of the skin over a large area and/or systemic effect. Furthermore, dressings containing medicinal products acting as ancillary to the dressing fall within Class III under Rule 13 (Table 3.11).

Table 3.11: Classification Rules as Per Rule 4 of MEDDEV 2.4/1

Rule 4	Examples
All non-invasive devices which come into contact with injured skin: • Are in Class I if they are intended to be used as a mechanical barrier, for compression or for absorption of exudates.	Wound dressings, such as absorbent pads, island dressings, cotton wool, wound strips, and gauze dressings to act as a barrier or to maintain the wound positionally or to absorb exudates from the wound.
• Are in Class IIb if they are intended to be used principally with wounds which have breached the dermis and can only heal by secondary intent	Are principally intended to be used with severe wounds that have substantially and extensively breached the dermis, and where the healing process can only be by secondary intent such as: • Dressings for chronic extensive ulcerated wounds. • Dressings for severe burns having breached the dermis and covering an extensive area. • Dressings for severe wounds. • Dressings incorporating means of augmenting tissue and providing a temporary skin substitute

• Are in Class IIa in all other cases, including devices principally intended to manage the micro-environment of a wound.	Have specific properties intended to assist the healing process by controlling the level of moisture at the wound during the healing process and to generally regulate the environment in terms of humidity and temperature, levels of oxygen and other gases and pH values or by influencing the process by other physical means. These devices may specify additional healing properties whilst not being intended for extensive wounds requiring healing by secondary intent. Adhesives for topical use. Polymer film dressings, hydrogel dressings and non-medicated impregnated gauze dressings.

- **Understanding the Practical Issues of Rule 4 Classification:** The products that fall within this rule are extremely claim sensitive, for example, a polymeric film dressing would fall in Class II A if the intended use is to manage the micro-environment of the wound and in Class I if its intended use is limited to retaining an invasive cannula at the wound site. Thus, the actual use of the device would determine the class, and it is impossible to say that a dressing is in a given class without knowing its intended use as defined by the manufacturer. So, for example, a claim that the device is interactive or active with respect to the wound healing process usually implies that the device is in Class II B.

It should also be noted that most dressings intended for a function/use that is in Class II A or II B will also perform functions that are in Class I, and as a result, such devices are nevertheless classed according to the intended use in the higher class.

For such devices incorporating medicines rule 13 should be consulted, and for devices incorporating animal tissues rule 17 should be consulted.

Special concepts defined by this rule include:

- **Breached Dermis:** whereby the wound exposes at least partly the subcutaneous tissues.
- **Secondary Intent:** whereby the wound heals by first being filled with granulation tissue. This is followed by the epithelium, which grows back over the granulation tissue and the wound contracts. This is in contrast to primary intent, which implies that the edges of the wound are close enough or pulled together, e.g., by suturing, to allow the wound to heal.

3.6.5. Rule 5: Devices Invasive in Body Orifices

General explanation of the rule: Rule 5 is intended to cover invasiveness with respect to the body orifices such as the ear, mouth, nose, eye, anus, urethra, and vagina. This must be considered separately from invasiveness that penetrates through a cut in the body surfaces, for example through surgical invasiveness. A further distinction must be made between invasiveness with respect to the less vulnerable anterior parts of the ear, mouth, and nose and the other anatomical sites that can be accessed through natural body orifices. The surgically created stoma, which for example allows the evacuation of urine or feces, should also be considered as a body orifice. Devices covered by this rule tend to be diagnostic and therapeutic instruments used in particular specialties (ENT, ophthalmology, dentistry, proctology, urology, and gynecology) (Table 3.12).

Table 3.12: Classification Rules as Per Rule 5 of MEDDEV 2.4/1

Rule 5	Examples
All invasive devices with respect to body orifices, other than surgically invasive devices and which are not intended for connection to an active medical device: • Are in Class I if they are intended for transient use. • Are in Class IIa if they are intended for short term use; except if they are used in the oral cavity as far as the pharynx, in an ear canal up to the eardrum or in a nasal cavity, in which case they are in Class I; • Are in Class IIb if they are intended for long term use; except if they are used in the oral cavity as far as the pharynx, in an ear canal up to the eardrum or in a nasal cavity and are not liable to be absorbed by the mucous membrane, in which case they are in Class IIa. All invasive devices with respect to body orifices, other than surgically invasive devices, intended for connection to an active medical device in Class IIa or a higher class, are in Class IIa.	Handheld mirrors used in dentistry to aid in dental diagnosis and surgery, dental impression materials, tubes used for pumping the stomach, impression tray, enema devices, examination gloves, and prostatic balloon dilation catheters. Contact lenses, urinary catheters, tracheal tubes, stents, vaginal pessaries and perineal re-education devices. Dressings for nose bleeds, dentures intended to be removed by the patient. Urethral stents Orthodontic wire, fixed dental prostheses, fissures sealants. Tracheostomy or tracheal tubes connected to a ventilator, blood oxygen analyzers placed under the eyelid, powered nasal irrigators, nasopharyngeal airways, some enteral feeding tubes, fiber optics in endoscopes connected to surgical lasers, suction catheters or tubes for stomach drainage, dental aspirator tips.

3.6.6. Rule 6: Surgically Invasive Devices for Transient Use

General explanation of the rule: This rule covers principally three major groups of devices, namely: (a) devices that are used to create a conduit through the skin such as needles; (b) surgical instruments such as scalpels and saws; (c) various types of catheters and/or suckers and so on (Table 3.13).

Table 3.13: Classification Rules as Per Rule 6 of MEDDEV 2.4/1

Rule 6	Examples
All surgically invasive[1] devices intended for transient use are in Class IIa unless they are: • Intended specifically to diagnose, monitor or correct a defect[2] of the heart or of the central circulatory system[1] through direct contact with these parts of the body, in which case they are in Class III[3] • Reusable surgical instruments[1], in which case they are in Class I[3] • Intended to supply energy in the form of ionizing radiation in which case they are in Class IIb • Intended to have a biological[5] effect or to be wholly or mainly absorbed[4] in which case they are in Class IIb • intended to administer medicines by means of a delivery system, if this is done in a manner that is potentially hazardous[6] considering of the mode of application, in which case they are Class IIb.	• Needles used for suturing, needles or syringes, lancets, suckers, single-use scalpels, single-use scalpel blades, support devices in ophthalmic surgery, staplers, surgical swabs, drill bits connected to active devices, surgical gloves, etchants, tester of artificial heart valves, heart valve sizers and holders, trial hip prosthesis heads or stems, swabs to sample exudates, single-use aortic punches (see note[2]) • Cardiovascular catheters (e.g., angioplasty balloon catheters), including related guidewires and dedicated[4] disposable cardiovascular surgical instruments, e.g., electrophysiological catheters, electrodes for electrophysiological diagnosis and ablation. Catheters containing or incorporating sealed radioisotopes, where the radioactive isotope as such is not intended to be released into the body, if used in the central circulatory system • Scalpels, scalpel handles, drill bits, saws, that are not intended for connection to an active device, and retractors forceps, excavators, and chisels. • Catheters containing or incorporating sealed radioisotopes, where the radioactive isotope as such is not intended to be released into the body, if used in the circulatory system, excluding the central circulatory system • Devices for repeated self-application where dosage levels and the nature of the medicinal product are critical, e.g., insulin pens.

- **Explanation of Special Concepts Applicable in Rule 6:**
- *Note 1:* Terms such as *"surgically invasive device," "central circulatory system"* and *"reusable surgical instruments"* are defined in Section I of Annex IX of the Directive. Surgical instruments that are connected to an active device are not considered to be "reusable surgical instruments."
- *Note 2:* The phrase "correct a defect" does not cover devices that are used as an accessory in heart surgery, for example clamps. The first indent of this rule does not apply to aortic punches and similar cutting instruments which perform a similar function to that of a scalpel.
- *Note 3:* This applies to surgical instruments which are not specifically intended for purposes described in the first indent of the table, and irrespective of the site of application, are 1) in class IIA if they are intended for single-use and 2) in class I if they are reusable.
- *Note 4: Dedicated* means that the intended purpose of the device is to diagnose, monitor or correct a defect of the heart or of the central circulatory system.
- *Note 5:* The phrase *"biological effect"* pertains to all materials and devices that have the potential to affect tissues following use in a surgically invasive procedure. A material is considered to have some kind of biological effect if it actively and intentionally induces, alters, or prevents a response from the tissues that is mediated by specific reactions at a molecular level. These kinds of devices may also be described as bioactive.

Wholly or mainly absorbed: Th term absorption refers to the degradation of a material within the body and the metabolic elimination of the resulting degradation products from the body.

- *Note 6:* The concept of *"potentially hazardous manner"* is related to the characteristics of the device and not the competence of the user.

3.6.7. Rule 7: Surgically Invasive Devices for Short-Term Us

General explanation of the rule: Devices covered by rule 7 are most those devices used in the context of surgery or post-operative care such as clamps or drains; infusion devise such as cannulas or needles and catheters of various types (Table 3.14).

Table 3.14: Classification Rules as Per Rule 7 of MEDDEV 2.4/1

Rule 7	**Examples**
All surgically invasive devices intended for short term use are in Class IIa unless they are intended:	Clamps, infusion cannulae, skin closure devices, temporary filling materials.
• Either specifically to diagnose, monitor or correct a defect[2] of the heart or of the central circulatory system through direct contact with these parts of the body, in which case they are in Class III;	• Tissue stabilizers[2] used in cardiac surgery • Cardiovascular catheters, cardiac output probes and temporary pacemaker leads. • Thoracic catheters intended to drain the heart, including the pericardium • Carotid artery shunts
• Or specifically for use in direct contact with the central nervous system, in which case they are in Class III;	• Neurological catheters, cortical electrodes
• Or to supply energy in the form of ionizing radiation in which case they are in Class IIb	• Brachytherapy devices
• Intended to have a biological effect or to be wholly or mainly absorbed in which case they are in Class III;	• Absorbable sutures and biological adhesives
• Or to undergo chemical change in the body, except if the devices are placed in the teeth, or to administer medicines[1], in which case they are Class IIb.	• Adhesives

- **Practical Issues of Rule 7 Classification:**
- *Note 1: "administer medicines"* – the administration of medicines is more than just channeling as it also implies storage and/or influencing the volume and rate of medicine being delivered. It should also be noted that implanted capsules for the slow release of medicines are actually medicines and not medical devices.
- *Note 2: "correct a defect"* – this phrase does not cover devices that are used as an accessory in heart surgery, for example, tissue stabilizers.

3.6.8. Rule 8: Surgically Invasive Devices for Long-Term Use and Implantable Devices

General explanation of the rule: devices falling under Rule 8 are mostly implants in the orthopedic, dental, ophthalmic, and cardiovascular fields, as well as soft tissue implants such as implants used in plastic surgery (Table 3.15).

Table 3.15: Classification Rules as Per Rule 8 of MEDDEV 2.4/1

Rule 8	Examples
All implantable devices and long-term surgically invasive devices are in Class IIb unless they are intended: • To be placed in the teeth[2], in which case they are in Class IIa • To be used in direct contact with the heart, the central circulatory system or the central nervous system, in which case they are Class III; • To have a biological effect or to be wholly or mainly absorbed, in which case they are in Class III; • Or to undergo chemical change[3] in the body, except if the devices are placed in the teeth, or to administer medicines, in which case they are in Class III.	• Prosthetic joint replacements, ligaments, shunts, stents, nails, plates, intra-ocular lenses, internal closure devices, tissue augmentation implants, infusion ports, peripheral vascular grafts, penile implants, non-absorbable sutures, bone cements and maxillo-facial implants, viscoelastic surgical devices intended specifically for ophthalmic anterior segment surgery[1] • Bridges, crowns, dental filling materials and pins, dental alloys, ceramics, and polymers. • Prosthetic heart valves, aneurysm clips, vascular prostheses, spinal stents, vascular stents, CNS electrodes, and cardiovascular sutures. • Permanent vena cava filters • Absorbable sutures, adhesives, and implantable devices claimed to be bioactive through the attachment of surface coatings such as phosphorylcholine. • Rechargeable non-active drug delivery systems.

- **Practical Issues of Classification for Rule 8:**
- *Note 1:* These products are implants because in normal conditions, a significant amount of the substance remains at the surgical site after the procedure. If, however, these devices contain animal tissues or derivatives of animal tissues, they are covered by rule 17.
- *Note 2: "to be placed in the teeth":* Implants without bioactive coatings intended to secure teeth or prostheses to the maxillary or mandibular bones are in Class II B following the general rule.
- *Note 3: "undergo chemical change":* The clause about chemical change under this rule does not apply to products such as bone cements where the chemical change takes place during the placement and does not continue in the long term.

3.6.9. Rule 9: Active Therapeutic Devices Intended to Administer or Exchange Energy

General explanation of the rule: Medical devices classified by Rule 8 are mostly electrical equipment used in surgery, such as lasers and surgical generators. There are, however, devices for specialized treatment such as radiation treatment. In addition, another category consists of stimulation devices, although not all of them can be considered as delivering dangerous levels of energy considering the tissue involved (Table 3.16).

Table 3.16: Classification Rules as Per Rule 9 of MEDDEV 2.4/1

Rule 9	Examples
All active therapeutic devices intended to administer, or exchange energy are in Class IIa	Electrical and/or magnetic and electromagnetic energy: Muscle stimulators and external bone growth stimulators, TENS devices and eye electromagnets, electrical acupuncture • Thermal energy: Cryosurgery equipment, heat exchangers, except the types described below • Mechanical energy: Powered dermatomes, powered drills, and dental handpieces. • Light: Phototherapy for skin treatment and for neonatal care • Sound: Hearing aids • Ultrasound: Equipment for physiotherapy
• Unless their characteristics are such that they may administer or exchange energy to and from the human body in a potentially hazardous way[1], taking account of the nature, the density and the site of application of the energy, in which case they are in Class IIb.	• Kinetic energy: Lung ventilators • Thermal energy: Incubators for babies, warming blankets, blood warmers, electrically powered heat exchangers, for instance, those used with patient's incapable of reacting, communicating, and/or who are without a sense of feeling • Electrical energy: High-frequency electrosurgical generators, and electrocautery equipment, including their electrodes, external pacemakers, external defibrillators, electroconvulsive therapy equipment • Coherent light: Surgical lasers • Ultrasound: surgical ultrasound devices • Ionizing radiation: Radioactive sources for afterloading therapy, therapeutic cyclotrons, linear accelerators, therapeutic X-ray sources.

All active devices intended to control and monitor the performance of active therapeutic devices in Class IIb or intended to influence directly the performance of such devices are in Class IIb.	External feedback systems for active therapeutic devices, after loading control devices.

- **Explanation of Special Concepts Pertaining to Rule 9:**
- *Note 1: "exchange energy to and from the human body in a potentially hazardous way":* The decision as to whether a medical device administers or exchanges energy to and from the human body in a potentially hazardous way should take various factors into consideration. The concept of "potentially hazardous" is dependent on the type of technology involved and the intended application of the device to the patient, and not on the measures adopted by the manufacturer in view of good design management. That being said, the manufacturer's obligation to comply with design requirements and solutions adopted (such as use of standards) exist independently from the classification system.

3.6.10. Rule 10: Active Devices for Diagnosis

General explanation of the rule: this rule covers primarily a whole range of widely used equipment in the field's ultrasound diagnosis and the capture of physiological signals as well as therapeutic and diagnostic radiology (Table 3.17).

Table 3.17: Classification Rules as Per Rule 10 of MEDDEV 2.4/1

Rule 10	Examples
Active devices intended for diagnosis are in Class IIa: • If they are intended to supply energy which will be absorbed by the human body, except for devices used to illuminate the patient's body, in the visible spectrum.	• Magnetic resonance equipment, pulp testers, evoked response stimulators, diagnostic ultrasound • Gamma cameras, positron emission tomography and single-photon emission computer tomography

• If they are intended to image *in vivo* distribution of radiopharmaceuticals. • If they are intended to allow direct diagnosis or monitoring of vital physiological processes[1]. unless they are specifically intended for monitoring of vital physiological parameters, where the nature of variations is such that it could result in immediate danger to the patient, for instance, variations in cardiac performance, respiration, activity of CNS, in which case they are in Class IIb. Active devices intended to emit ionizing radiation and intended for diagnostic and therapeutic interventional radiology[3] including devices which control or monitor[4] such devices, or which directly influence their performance, are in Class II B.	• Electrocardiographs, electroencephalographs, cardioscopes with or without pacing pulse indicators[2] • Electronic thermometers • Electronic stethoscopes • Electronic blood pressure measuring equipment • Intensive care monitoring and alarm devices (for, e.g., blood pressure, temperature, oxygen saturation), biological sensors, blood gas analyzers used in open-heart surgery, cardioscopes, and apnea monitors, including apnea monitors in home care • Diagnostic X-ray sources.

- **Explanation of Special Concepts Pertaining to Rule 10:**
- *Note 1:* "vital physiological processes": These vital physiological processes and parameters include, for example; respiration, heart rate, cerebral functions, blood gases, blood pressure, and body temperature. Medical devices intended to be used for continuous surveillance of vital physiological processes in anesthesia, intensive care or emergency care are in Class IIB, whilst medical devices intended to be used to obtain readings of vital physiological signals in routine check-ups and in self-monitoring are in Class IIA. A thermal imaging device intended to monitor blood flow is not considered to be a temperature measuring device.
- *Note 2:* "With or without pacing pulse indicators": This is applicable for devices specifically intended to monitor AIMDs and which fall under the AIMD Directive.
- *Note 3:* Therapeutic interventional radiology: This refers to diagnosis being carried out during surgical procedures.
- *Note 4:* "including devices which control or monitor": This refers

to active devices for the control, monitoring or influencing and not to the subsequent processing, recording or viewing of the resulting image.

3.6.11. Rule 11: Active Devices to Administer, Remove Medicines and Other Substances to or From the Body

General explanation of the rule: this rule is intended to cover primarily drug delivery systems and anesthesia equipment (Table 3.18).

Table 3.18: Classification Rules as Per Rule 11 of MEDDEV 2.4/1

Rule 11	Examples
All active devices intended to administer and/or remove medicines, body-liquids or other substances to or from the body are in Class IIa. • Unless this is done in a manner: that is potentially hazardous, taking account of the nature of the substances involved, of the part of the body concerned and of the mode of application, in which case they are in Class IIb.	• Suction equipment, feeding pumps. • Jet injectors for vaccination • Nebulizers to be used on conscious and spontaneously breathing patients where failure to deliver the appropriate dosage characteristics is not potentially hazardous • Infusion pumps, ventilators, anesthesia machines, anesthetic vaporizers, dialysis equipment, blood pumps for heart-lung machines, hyperbaric chambers, pressure regulators for medical gases, medical gas mixers, moisture exchangers in breathing circuits if used on unconscious or non-spontaneously breathing patients • Nebulizers where the failure to deliver the appropriate dosage characteristics could be hazardous.

3.6.12. Rule 12: All Other Active Devices

General explanation of the rule: this forms a fallback rule to cover all active devices not covered by previous rules (Table 3.19).

Table 3.19: Classification Rules as Per Rule 12 of MEDDEV 2.4/1

Rule 12	Examples
All other active devices are in Class I	Active diagnostic devices intended to illuminate the patient's body in the visible spectrum such as examination lights or to optically view the body such as surgical microscopes. • Devices intended in general for external patient support (e.g., hospital beds, patient hoists, walking aids, wheelchairs, stretchers, dental patient chairs). • Active diagnostic devices intended for thermography. • Active devices intended for recording, processing or viewing of diagnostic images. • dental curing lights.

The special rules are discussed in the following subsections.

3.6.13. Rule 13: Devices Incorporating a Medicinal Substance

General explanation of the rule: This rule is intended to cover combination devices that contain a medicinal substance incorporated into the device. This substance is usually present for the purpose of assisting the functioning of that device. This rule, however, does not cover those devices incorporating substances which under other circumstances may be considered as medicinal substances. It does however cover that which is incorporated into the device exclusively for the purpose of maintaining certain characteristics of the device and which are not liable to act on the body. A common example includes agents for the preservation of solutions for contact lenses. The primary function of the device does not rely on the pharmacological effect of the medicine as if the latter is true, then the product would be classified as being a medicine rather than a device. In that case, it will not be covered by the MEDDEV 2.4/1 directive (Table 3.20).

Table 3.20: Classification Rules as Per Rule 13 of MEDDEV 2.4/1

Rule 13	Examples
All devices incorporating, as an integral part[1], a substance which, if used separately, can be considered to be a medicinal product as defined in Article 1 of the Directive 65/65/EEC, and which is liable to act on the human body with action ancillary to that of the devices, are in Class III.	• Antibiotic bone cements, condoms with spermicide, heparin-coated catheters, endodontic materials with antibiotics. • Ophthalmic irrigation solutions principally intended for irrigation, which contain components which support the metabolism of the endothelial cells of the cornea • Dressings incorporating an antimicrobial agent where the purpose of such an agent is to provide ancillary action on the wound

- *Note 1:* Integral part: Integral part means that the device and the medicinal substance forms one physical unit.

3.6.14. Rule 14: Devices Used for Contraception or Prevention of Sexually Transmitted Diseases

General explanation of the rule: the intended uses defined by this rule relates to special cases of human vulnerability that cannot be covered by the normal criteria of time, invasiveness, and organic function. While the rule does cover two very different device applications, some devices may perform both functions. A common example of this would be condoms. It should also be noted that devices intended to prevent the sexual transmission of HIV are also covered by this rule (Table 3.21).

Table 3.21: Classification Rules as Per Rule 14 of MEDDEV 2.4/1

Rule 14	Examples
All devices used for contraception or the prevention of the transmission of sexually transmitted diseases are in Class IIb. unless they are implantable or long-term invasive devices, in which case they are in Class III.	• Condoms, contraceptive diaphragms • Contraceptive intrauterine devices (IUDs')1

- *Note 1:* Contraceptive intrauterine devices (IUDs): Those intrauterine contraceptives whose primary purpose is to release progestogens are not medical devices.

3.6.15. Rule 15: Specific Disinfecting, Cleaning, and Rinsing Devices

General explanation of the rule: this rule is mainly intended to cover various contact lens fluids as well as those substances used principally in a medical environment to disinfect medical devices (Table 3.22).

Table 3.22: Classification Rules as Per Rule 15 of MEDDEV 2.4/1

Rule 15	Examples
All devices intended specifically to be used for disinfecting, cleaning, rinsing, or, when appropriate hydrating contact lenses are in Class IIb. All devices intended specifically to be used for disinfecting medical devices are in Class IIa. This rule does not apply to products that are intended to clean medical devices other than contact lenses by means of physical action[1]	• Contact lens solutions, comfort solutions. • Disinfectants specifically intended for instance for endoscopes or hemodialysis equipment, sterilizers specifically intended to sterilize medical devices in a medical environment, and washer-disinfectors. • Cleaners which disinfect prosthetic dentures.

- **Practical Issues of Rule 15 Classification:**
- *Note 1:* This rule does not apply to mechanical means of cleaning of devices, such as brushes and ultrasound. Such products will only fall under this directive if they are specifically intended for use with medical devices.

3.6.16. Rule 16: Non-Active Devices to Record X-Ray Diagnostic Images (Table 3.23)

Table 3.23: Classification Rules as Per Rule 16 of MEDDEV 2.4/1

Rule 16	Examples
Non-active devices specifically intended for recording of X-ray diagnostic images are in Class II a.	X-ray films, photostimulable phosphor plates

- *Note:* this rule refers to primary recording media such as X-ray films and not to media used for subsequent reproduction.

3.6.17. Rule 17: Devices Utilizing Animal Tissue or Derivatives

General explanation of the rule: Rule 17 covers those devices that contain animal tissues, or which are made of animal tissues that have been rendered non-viable. It also considers those derivatives from such tissues where there is no longer any capacity for cellular metabolic activity (i.e.: non-viable tissues). It should be noted that those medical devices containing non-inactivated animal tissues and/or any human tissues or derivatives are excluded from the scope of this Directive. It should be further noted that the manufacture of some medical devices may use industrial raw materials, and this may result in the presence of small amounts of tallow or tallow derivatives which are not considered as derivatives of animal tissues for the purpose of this rule. In this case, this rule does not apply (Table 3.24).

Table 3.24: Classification Rules as Per Rule 17 of MEDDEV 2.4/1

Rule 17	Examples
All devices manufactured utilizing animal tissues or derivatives[1] rendered non-viable are Class III except where such devices are intended to come into contact with intact skin[2] only.	Biological heart valves, porcine xenograft dressings, catgut sutures, implants, and dressings made from collagen.

Practical classification issues: those medical devices made of non-viable animal tissue that come into contact with intact skin only are in Class I in accordance to rule 1.

- *Note 1:* Animal tissues or derivatives: Derivatives are defined as those products that are processed from animal tissues and exclude substances such as milk, silk, beeswax, hair, lanolin.
- *Note 2:* Intact skin: Intact skin includes the skin around an established stoma unless however, this skin is breached.

3.6.18. Rule 18: Blood Bags

General explanation of the rule: This is a special rule that covers only blood bags (Table 3.25).

Table 3.25: Classification Rules as Per Rule 18 of MEDDEV 2.4/1

Rule 18	Examples
By derogation from other rules, blood bags are in Class IIb	Blood bags (including those containing or coated with an anticoagulant). Where blood bags have a function greater than for storing purposes and include systems for preservation other than anticoagulants then other rules (e.g., rule 13) may apply.

Chapter 4

Quality Management System

CONTENTS

4.1. CONFORMITY ASSESSMENTS

For a medical device manufacturer, one of the biggest requirements is being able to demonstrate compliance to relevant regulatory authorities. The EU regulations give manufacturers the ability to choose which conformity assessment route they would like to use, but this is dependent on the class of the device itself. Choosing the right conformity assessment procedure is important to ensure a manufacturer gets approval without wasting precious resources or delaying the launch of the product.

4.2. UNDERSTANDING CONFORMITY ASSESSMENT UNDER THE MDD

Article 11 of the MDD lists the choices a manufacturer has when it comes to choosing a devices conformity assessment route. The reason for the different choices is to provide flexibility as each manufacturer is different, and what works well for one manufacturer might not work well for another. It should be noted that all conformity assessment procedures will require a manufacturer to have technical documentation pertaining to the device, and this will include information about how it confirms to meet the essential requirements. The notified body will perform a technical documentation review of the submitted assessment. It is important to remember that the notified body won't issue an EC certificate without the relevant technical documentation in place and failing to obtain an EC certificate means a manufacturer is unable to place the CE mark on the device and ultimately sell within the EU.

There are six conformity assessment options under the MDD:

- **Annex II:** Full quality assurance system.
- **Annex III:** EC type-examination.
- **Annex IV:** EC verification.
- **Annex V:** EC Declaration of conformity-production quality assurance.
- **Annex VI:** EC Declaration of conformity-product quality assurance.
- **Annex VII:** EC Declaration of conformity.

Article 11 of the MDD is highlighted below for the manufacturer:

Article 11 of the MDD 93/42/EEC: Conformity Assessment Procedures (As Taken from the Directive)

1. In the case of devices falling within Class III, other than devices which are custom-made or intended for clinical investigations, the manufacturer shall, in order to affix the CE marking, either:

i. follow the procedure relating to the EC declaration of conformity set out in Annex II (full quality assurance); or

ii. Follow the procedure relating to the EC type-examination set out in Annex III, coupled with:

a. The procedure relating to the EC verification set out in Annex IV; or

b. The procedure relating to the EC declaration of conformity set out in Annex V (production quality assurance).

2. In the case of devices falling within Class IIa, other than devices which are custom-made or intended for clinical investigations, the manufacturer shall, in order to affix the CE marking, follow the procedure relating to the EC declaration of conformity set out in Annex VII, coupled with either:

i. The procedure relating to the EC verification set out in Annex IV;

ii. The procedure relating to the EC declaration of conformity set out in Annex V (production quality assurance); or

iii. The procedure relating to the EC declaration of conformity set out in Annex VI (product quality assurance).

Instead of applying these procedures, the manufacturer may also follow the procedure referred to in paragraph 3 (a).

3. In the case of devices falling within Class IIb, other than devices which are custom-made or intended for clinical investigations, the manufacturer shall, in order to affix the CE marking, either:

i. follow the procedure relating to the EC declaration of conformity set out in Annex II (full quality assurance); in this case, point 4 of Annex II is not applicable; or

ii. Follow the procedure relating to the EC type-examination set out in Annex III, coupled with:

a. The procedure relating to the EC verification set out in Annex IV;

b. The procedure relating to the EC declaration of conformity set out in Annex V (production quality assurance); or

c. The procedure relating to the EC declaration of conformity set out in Annex VI (product quality assurance).

4. The Commission shall, no later than five years from the date of implementation of this Directive, submit a report to the Council on the operation of the provisions referred to in Article 10 (1), Article 15 (1), in particular in respect of Class I and Class IIa devices, and on the operation of the provisions referred to in Annex II, Section 4.3 second and third subparagraphs and in Annex III, Section 5 second and third subparagraphs to this Directive, accompanied, if necessary, by appropriate proposals.

5. In the case of devices falling within Class I, other than devices which are custom-made or intended for clinical investigations, the manufacturer shall, in order to affix the CE marking, follow the procedure referred to in Annex VII and draw up the EC declaration of conformity required before placing the device on the market.

6. In the case of custom-made devices, the manufacturer shall follow the procedure referred to in Annex VIII and draw up the statement set out in that Annex before placing each device on the market. Member States may require that the manufacturer shall submit to the competent authority a list of such devices which have been put into service in their territory.

7. During the conformity assessment procedure for a device, the manufacturer and/or the notified body shall take account of the results of any assessment and verification operations which, where appropriate, have been carried out in accordance with this Directive at an intermediate stage of manufacture.

8. The manufacturer may instruct his authorized representative to initiate the procedures provided for in Annexes III, IV, VII, and VIII.

9. Where the conformity assessment procedure involves the intervention of a notified body, the manufacturer, or his authorized representative, may apply to a body of his choice within the framework of the tasks for which the body has been notified.

10. The notified body may require, where duly justified, any information or data, which is necessary for establishing and maintaining the attestation of conformity in view of the chosen procedure.

11. Decisions taken by the notified bodies in accordance with Annexes II, III, V, and VI shall be valid for a maximum of five years and maybe extended on application, made at a time agreed in the contract signed by both parties, for further periods of a maximum length of five years.

12. The records and correspondence relating to the procedures referred to in paragraphs 1 to 6 shall be in an official language of the Member State in which the procedures are carried out and/or in another Community language acceptable to the notified body.

13. By derogation from paragraphs 1 to 6, the competent authorities may authorize, on duly justified request, the placing on the market and putting into service, within the territory of the Member State concerned, of individual devices for which the procedures referred to in paragraphs 1 to 6 have not been carried out and the use of which is in the interest of protection of health.

14. The measures designed to amend non-essential elements of this Directive, by supplementing it, relating to the means by which, in the light of technical progress and considering the intended users of the devices concerned, the information laid down in Annex I Section 13.1 may be set out, shall be adopted in accordance with the regulatory procedure with scrutiny referred to in Article 7(3).

The six conformity assessment options are explained in detail below:

- **Annex II: Full Quality Assurance System:** The full quality assurance system conformity assessment is one of the more common assessment used and includes a full review of the manufacturers entire quality management system (QMS). This include design, manufacturing, and final inspection. Regardless of whether a device falls into class III or not, when a manufacturer undergoes assessment, the technical documentation will still be reviewed. If the product is a class III device and annex II is followed for conformity assessment, the manufacturer will need to include a design dossier for the review. This will need to show conformity to the Essential Requirements.

Annex II from the MDD is highlighted below:

Annex II of the MDD 93/42/EEC: EC Declaration of Conformity (Full Quality Assurance System)	
1.	The manufacturer must ensure application of the quality system approved for the design, manufacture, and final inspection of the products concerned, as specified in Section 3 and is subject to audit as laid down in Sections 3.3 and 4 and to Community surveillance as specified in Section 5.
2.	The EC declaration of conformity is the procedure whereby the manufacturer who fulfills the obligations imposed by Section 1 ensures and declares that the products concerned meet the provisions of this Directive which apply to them. The manufacturer must affix the CE marking in accordance with Article 17 and draw up a written declaration of conformity. This declaration must cover one or more medical devices manufactured, clearly identified by means of product name, product code or other unambiguous reference, and must be kept by the manufacturer.
3.	Quality system:
3.1.	The manufacturer must lodge an application for assessment of his quality system with a notified body. The application must include:
i.	The name and address of the manufacturer and any additional manufacturing site covered by the quality system.
ii.	All the relevant information on the product or product category covered by the procedure.
iii.	A written declaration that no application has been lodged with any other notified body for the same product-related quality system.
iv.	The documentation on the quality system.
v.	An undertaking by the manufacturer to fulfill the obligations imposed by the quality system approved.
vi.	An undertaking by the manufacturer to keep the approved quality system adequate and efficacious.
vii.	An undertaking by the manufacturer to institute and keep up to date a systematic procedure to review experience gained from devices in the post-production phase, including the provisions referred to in Annex X, and to implement appropriate means to apply any necessary corrective action. This undertaking must include an obligation for the manufacturer to notify the competent authorities of the following incidents immediately on learning of them:

a. Any malfunction or deterioration in the characteristics and/or performance of a device, as well as any inadequacy in the instructions for use which might lead to or might have led to the death of a patient or user or to a serious deterioration in his state of health;

b. Any technical or medical reason connected with the characteristics or performance of a device leading for the reasons referred to in subparagraph (i) to systematic recall of devices of the same type by the manufacturer.

3.2. Application of the quality system must ensure that the products conform to the provisions of this Directive which apply to them at every stage, from design to final inspection. All the elements, requirements, and provisions adopted by the manufacturer for his quality system must be documented in a systematic and orderly manner in the form of written policies and procedures such as quality programs, quality plans, quality manuals, and quality records.

It shall include in particular the corresponding documentation, data, and records arising from the procedures referred to in point (c).

It shall include in particular an adequate description of:

i. The manufacturer's quality objectives;

ii. The organization of the business and in particular:

a. The organizational structures, the responsibilities of the managerial staff and their organizational authority where quality of design and manufacture of the products is concerned.

b. The methods of monitoring the efficient operation of the quality system and in particular its ability to achieve the desired quality of design and of product, including control of products which fail to conform.

c. Where the design, manufacture, and/or final inspection and testing of the products, or elements thereof, is carried out by a third party, the methods of monitoring the efficient operation of the quality system and in particular the type and extent of control applied to the third party;

iii. The procedures for monitoring and verifying the design of the products, including the corresponding documentation, and in particular:

a. A general description of the product, including any variants planned, and its intended use(s).

b. The design specifications, including the standards which will be applied and the results of the risk analysis, and also a description of the solutions adopted to fulfill the essential requirements which apply to the products if the standards referred to in Article 5 are not applied in full.

c. The techniques used to control and verify the design and the processes and systematic measures which will be used when the products are being designed.

d. If the device is to be connected to other device(s) in order to operate as intended, proof must be provided that it conforms to the essential requirements when connected to any such device(s) having the characteristics specified by the manufacturer.

e. A statement indicating whether or not the device incorporates, as an integral part, a substance or a human blood derivative referred to in section 7.4 of Annex I and the data on the tests conducted in this connection required to assess the safety, quality, and usefulness of that substance or human blood derivative, taking account of the intended purpose of the device.

f. A statement indicating whether or not the device is manufactured utilizing tissues of animal origin as referred to in Commission Directive 2003/32/EC (1).

g. The solutions adopted as referred to in Annex I, Chapter I, Section 2.

h. The pre-clinical evaluation.

i. The clinical evaluation referred to in Annex X.

j. The draft label and, where appropriate, instructions for use.

iv. The inspection and quality assurance techniques at the manufacturing stage and in particular:

a. The processes and procedures which will be used, particularly as regards sterilization, purchasing, and the relevant documents.

b. The product identification procedures drawn up and kept up to date from drawings, specifications or other relevant documents at every stage of manufacture;

v. The appropriate tests and trials which will be carried out before, during, and after manufacture, the frequency with which they will take place, and the test equipment used; it must be possible to trace back the calibration of the test equipment adequately.

3.3. The notified body must audit the quality system to determine whether it meets the requirements referred to in Section 3.2. It must presume that quality systems that implement the relevant harmonized standards conform to these requirements. The assessment team must include at least one member with past experience of assessments of the technology concerned. The assessment procedure must include an assessment, on a representative basis, of the documentation of the design of the product(s) concerned, an inspection on the manufacturer's premises and, in duly substantiated cases, on the premises of the manufacturer's suppliers and/or subcontractors to inspect the manufacturing processes.

The decision is notified to the manufacturer. It must contain the conclusions of the inspection and a reasoned assessment.

3.4. The manufacturer must inform the notified body which approved the quality system of any plan for substantial changes to the quality system or the product range covered. The notified body must assess the changes proposed and verify whether after these changes the quality system still meets the requirements referred to in Section 3.2. It must notify the manufacturer of its decision. This decision must contain the conclusions of the inspection and a reasoned assessment.

4. Examination of the design of the product

4.1. In addition to the obligations imposed by Section 3, the manufacturer must lodge with the notified body an application for examination of the design dossier relating to the product which he plans to manufacture and which falls into the category referred to in Section 3.1.

4.2. The application must describe the design, manufacture, and performances of the product in question. It must include the documents needed to assess whether the product conforms to the requirements of this Directive, as referred to in Section 3.2 (c).

4.3. The notified body must examine the application and, if the product conforms to the relevant provisions of this Directive, issue the application with an EC design-examination certificate. The notified body may require the application to be completed by further tests or proof to allow assessment of conformity with the requirements of the Directive. The certificate must contain the conclusions of the examination, the conditions of validity, the data needed for identification of the approved design, where appropriate, a description of the intended purpose of the product.

In the case of devices referred to in Annex I, Section 7.4, second paragraph, the notified body shall, as regards the aspects referred to in that section, consult one of the competent authorities designated by the Member States in accordance with Directive 2001/83/EC or the EMEA before taking a decision. The opinion of the competent national authority or the EMEA must be drawn up within 210 days after receipt of valid documentation. The scientific opinion of the competent national authority or the EMEA must be included in the documentation concerning the device. The notified body will give due consideration to the views expressed in this consultation when making its decision. It will convey its final decision to the competent body concerned.

In the case of devices referred to in Annex I, Section 7.4, third paragraph, the scientific opinion of the EMEA must be included in the documentation concerning the device. The opinion of the EMEA must be drawn up within 210 days after receipt of valid documentation. The notified body will give due consideration to the opinion of the EMEA when making its decision. The notified body may not deliver the certificate if the EMEA's scientific opinion is unfavorable. It will convey its final decision to the EMEA.

In the case of devices manufactured utilizing tissues of animal origin as referred to in Directive 2003/32/EC, the notified body must follow the procedures referred to in that Directive.

4.4. Changes to the approved design must receive further approval from the notified body which issued the EC design-examination certificate wherever the changes could affect conformity with the essential requirements of the Directive or with the conditions prescribed for use of the product. The applicant shall inform the notified body which issued the EC design-examination certificate of any such changes made to the approved design. This additional approval must take the form of a supplement to the EC design examination certificate.

5. Surveillance

5.1. The aim of surveillance is to ensure that the manufacturer duly fulfills the obligations imposed by the approved quality system.

5.2. The manufacturer must authorize the notified body to carry out all the necessary inspections and supply it with all relevant information, in particular:

i. The documentation on the quality system.

ii. The data stipulated in the part of the quality system relating to design, such as the results of analyzes, calculations, tests, the solutions adopted as referred to in Annex I, Chapter I, Section 2, pre-clinical, and clinical evaluation, post-market clinical follow-up plan and the results of the post-market clinical follow-up, if applicable, etc.

iii. The data stipulated in the part of the quality system relating to manufacture, such as inspection reports and test data, calibration data, qualification reports of the personnel concerned, etc.

5.3. The notified body must periodically carry out appropriate inspections and assessments to make sure that the manufacturer applies the approved quality system and must supply the manufacturer with an assessment report.

5.4. In addition, the notified body may pay unannounced visits to the manufacturer. At the time of such visits, the notified body may, where necessary, carry out or ask for tests in order to check that the quality system is working properly. It must provide the manufacturer with an inspection report and, if a test has been carried out, with a test report.

6. Administrative provisions

6.1. The manufacturer or his authorized representative must, for a period ending at least five years, and in the case of implantable devices at least 15 years, after the last product has been manufactured, keep at the disposal of the national authorities:

i. The declaration of conformity.

ii. The documentation referred to in the fourth indent of Section 3.1 and in particular the documentation, data, and records referred to in the second paragraph of Section 3.2.

iii. The changes referred to in Section 3.4.

iv. The documentation referred to in Section 4.2, and

v. The decisions and reports from the notified body as referred to in Sections 3.3, 4.3, 4.4, 5.3, and 5.4.

7. Application to devices in Classes IIa and IIb.

7.1. In line with Article 11(2) and (3), this Annex may apply to products in Classes IIa and IIb. Section 4, however, does not apply.

7.2. For devices in Class IIa, the notified body shall assess, as part of the assessment in Section 3.3, the technical documentation as described in Section 3.2(c) for at least one representative sample for each device subcategory for compliance with the provisions of this Directive.

7.3. For devices in Class IIb, the notified body shall assess, as part of the assessment in Section 3.3, the technical documentation as described in Section 3.2(c) for at least one representative sample for each generic device group for compliance with the provisions of this Directive.

7.4.	In choosing representative sample(s) the notified body shall take into account the novelty of the technology, similarities in design, technology, manufacturing, and sterilization methods, the intended use and the results of any previous relevant assessments (e.g., with regard to physical, chemical or biological properties) that have been carried out in accordance with this Directive. The notified body shall document and keep available to the competent authority its rationale for the sample(s) taken.
7.5.	Further samples shall be assessed by the notified body as part of the surveillance assessment referred to in Section 5.
8.	Application to the devices referred to Article 1(4a)
Upon completing the manufacture of each batch of devices referred to in Article 1(4a), the manufacturer shall inform the notified body of the release of the batch of devices and send to it the official certificate concerning the release of the batch of human blood derivative used in the device, issued by a State laboratory or a laboratory designated for that purpose by a Member State in accordance with Article 114(2) of Directive 2001/83/EC.	

1. **Annex III: EC Type Examination:** The EC type-examination conformity assessment, the notified body will examine a certain representative sample of the device to show conformity to the Essential Requirements. These assessments look similar to that of the design dossier assessments. In this type of assessment, the notified body will perform an off-site, in-depth review of the technical requirements.

Annex III from the MDD is highlighted below:

Annex III of the MDD 93/42/EEC: EC Type Examination	
1.	EC type-examination is the procedure whereby a notified body ascertains and certifies that a representative sample of the production covered fulfills the relevant provisions of this Directive.
2.	The application includes:
i.	The name and address of the manufacturer and the name and address of the authorized representative if the application is lodged by the representative.
ii.	The documentation described in Section 3 needed to assess the conformity of the representative sample of the production in question, hereinafter referred to as the 'type,' with the requirements of this Directive. The applicant must make a 'type' available to the notified body. The notified body may request other samples as necessary.
iii.	A written declaration that no application has been lodged with any other notified body for the same type.

3. The documentation must allow an understanding of the design, the manufacture and the performances of the product and must contain the following items in particular:

i. A general description of the type, including any variants planned, and its intended use(s).

ii. Design drawings, methods of manufacture envisaged, in particular as regards sterilization, and diagrams of components, sub-assemblies, circuits, etc.

iii. The descriptions and explanations necessary to understand the above-mentioned drawings and diagrams and the operation of the product.

iv. A list of the standards referred to in Article 5, applied in full or in part, and descriptions of the solutions adopted to meet the essential requirements if the standards referred to in Article 5 have not been applied in full.

v. The results of the design calculations, risk analysis, investigations, technical tests, etc., carried out.

vi. A statement indicating whether or not the device incorporates, as an integral part, a substance, or human blood derivative, referred to in Section 7.4 of Annex I, and the data on the tests conducted in this connection which are required to assess the safety, quality, and usefulness of that substance, or human blood derivative, taking account of the intended purpose of the device.

vii. A statement indicating whether or not the device is manufactured utilizing tissues of animal origin as referred to in Directive 2003/32/EC.

viii. The solutions adopted as referred to in Annex I, Chapter I, Section 2.

ix. The preclinical evaluation.

x. The clinical evaluation referred to in Annex X.

xi. The draft label and, where appropriate, instructions for use.

4. The notified body must:

4.1. Examine and assess the documentation and verify that the type has been manufactured in conformity with that documentation; it must also record the items designed in conformity with the applicable provisions of the standards referred to in Article 5, as well as the items not designed on the basis of the relevant provisions of the above-mentioned standards;

4.2. Carry out or arrange for the appropriate inspections and the tests necessary to verify whether the solutions adopted by the manufacturer meet the essential requirements of this Directive if the standards referred to in Article 5 have not been applied; if the device is to be connected to other device(s) in order to operate as intended, proof must be provided that it conforms to the essential requirements when connected to any such device(s) having the characteristics specified by the manufacturer;

4.3. Carry out or arrange for the appropriate inspections and the tests necessary to verify whether, if the manufacturer has chosen to apply the relevant standards, these have actually been applied;

4.4. Agree with the applicant on the place where the necessary inspections and tests will be carried out.

5. If the type conforms to the provisions of this Directive, the notified body issues the applicant with an EC type-examination certificate. The certificate must contain the name and address of the manufacturer, the conclusions of the inspection, the conditions of validity and the data needed for identification of the type approved. The relevant parts of the documentation must be annexed to the certificate and a copy kept by the notified body.

In the case of devices referred to in Annex I, Section 7.4, second paragraph, the notified body shall, as regards the aspects referred to in that section, consult one of the authorities designated by the Member States in accordance with Directive 2001/83/EC or the EMEA before taking a decision. The opinion of the competent national authority or the EMEA must be drawn up within 210 days after receipt of valid documentation. The scientific opinion of the competent national authority or the EMEA must be included in the documentation concerning the device. The notified body will give due consideration to the views expressed in this consultation when making its decision. It will convey its final decision to the competent body concerned.

In the case of devices referred to in Annex I, Section 7.4, third paragraph, the scientific opinion of the EMEA must be included in the documentation concerning the device. The opinion of the EMEA must be drawn up within 210 days after receipt of valid documentation. The notified body will give due consideration to the opinion of the EMEA when making its decision. The notified body may not deliver the certificate if the EMEA's scientific opinion is unfavorable. It will convey its final decision to the EMEA.

In the case of devices manufactured utilizing tissues of animal origin as referred to in Directive 2003/32/EC, the notified body must follow the procedures referred to in that Directive.

6. The applicant must inform the notified body which issued the EC type-examination certificate of any significant change made to the approved product.

Changes to the approved product must receive further approval from the notified body which issued the EC type-examination certificate wherever the changes may affect conformity with the essential requirements or with the conditions prescribed for the use of the product. This new approval must, where appropriate, take the form of a supplement to the initial EC type-examination certificate.

7. Administrative provisions

7.1. Other notified bodies may obtain a copy of the EC type-examination certificates and/or the supplements thereto. The Annexes to the certificates must be made available to other notified bodies on reasoned application after the manufacturer has been informed.

7.2. The manufacturer or his authorized representative must keep with the technical documentation copies of EC type-examination certificates and their additions for a period ending at least five years after the last device has been manufactured. In the case of implantable devices, the period shall be at least 15 years after the last product has been manufactured.

- **Annex IV: EC Verification:** In an EC verification type of conformity assessment, the notified body will examine and test the product to verify its conformity. The notified body will either test and assess each individual product, or they will use a statistical sampling method to make sure that they test and assess an adequate sample size to determine conformity. If a manufacturer passes, then the notified body will certify each product or each batch. The certificate issued will only apply to that product or batch that is examined. It should be noted that while this assessment does not have QMS requirements, it does have requirements for the manufacturers post-market surveillance system. It also requires a manufacturer to provide technical documentation for those devices that demonstrate conformity to the Essential Requirements.

Annex IV from the MDD is highlighted below:

Annex IV of the MDD 93/42/EEC: EC Verification

1. EC verification is the procedure whereby the manufacturer or his authorized representative ensures and declares that the products which have been subject to the procedure set out in Section 4 conform to the type described in the EC type-examination certificate and meet the requirements of this Directive which apply to them.
2. The manufacturer must take all the measures necessary to ensure that the manufacturing process produces products which conform to the type described in the EC type-examination certificate and to the requirements of the Directive which apply to them. Before the start of manufacture, the manufacturer must prepare documents defining the manufacturing process, in particular as regards sterilization where necessary, together with all the routine, pre-established provisions to be implemented to ensure homogeneous production and, where appropriate, conformity of the products with the type described in the EC type-examination certificate and with the requirements of this Directive which apply to them. The manufacturer must affix the CE marking in accordance with Article 17 and draw up a declaration of conformity.

In addition, for products placed on the market in sterile condition, and only for those aspects of the manufacturing process designed to secure and maintain sterility, the manufacturer must apply the provisions of Annex V, Sections 3 and 4.

3. The manufacturer must undertake to institute and keep up to date a systematic procedure to review experience gained from devices in the post-production phase, including the provisions referred to in Annex X, and to implement appropriate means to apply any necessary corrective action. This undertaking must include an obligation for the manufacturer to notify the competent authorities of the following incidents immediately on learning of them:

i. Any malfunction or deterioration in the characteristics and/or performance of a device, as well as any inadequacy in the labeling or the instructions for use which might lead to or might have led to the death of a patient or user or to a serious deterioration in his state of health;

ii. Any technical or medical reason connected with the characteristics or performance of a device for the reasons referred to in subparagraph (i) leading to systematic recall of devices of the same type by the manufacturer.

4. The notified body must carry out the appropriate examinations and tests in order to verify the conformity of the product with the requirements of the Directive either by examining and testing every product as specified in Section 5 or by examining and testing products on a statistical basis as specified in Section 6, as the manufacturer decides. The aforementioned checks do not apply to those aspects of the manufacturing process designed to secure sterility.

5. Verification by examination and testing of every product

5.1. Every product is examined individually and the appropriate tests defined in the relevant standard(s) referred to in Article 5 or equivalent tests must be carried out in order to verify, where appropriate, the conformity of the products with the EC type described in the type-examination certificate and with the requirements of the Directive which apply to them.

5.2. The notified body must affix, or have affixed its identification number to each approved product and must draw up a written certificate of conformity relating to the tests carried out.

6. Statistical verification

6.1. The manufacturer must present the manufactured products in the form of homogeneous batches.

6.2. A random sample is taken from each batch. The products which make up the sample are examined individually and the appropriate tests defined in the relevant standard(s) referred to in Article 5 or equivalent tests must be carried out to verify, where appropriate, the conformity of the products with the type described in the EC type-examination certificate and with the requirements of the Directive which apply to them in order to determine whether to accept or reject the batch.

6.3. Statistical control of products will be based on attributes and/or variables, entailing sampling schemes with operational characteristics which ensure a high level of safety and performance according to the state of the art. The sampling schemes will be established by the harmonized standards referred to in Article 5, taking account of the specific nature of the product categories in question.

6.4. If the batch is accepted, the notified body affixes or has affixed its identification number to each product and draws up a written certificate of conformity relating to the tests carried out. All products in the batch may be put on the market except any in the sample which failed to conform.

If a batch is rejected, the competent notified body must take appropriate measures to prevent the batch from being placed on the market. In the event of frequent rejection of batches, the notified body may suspend the statistical verification. The manufacturer may, on the responsibility of the notified body, affix the notified body's identification number during the manufacturing process.

7. Administrative provisions

The manufacturer or his authorized representative must, for a period ending at least five years, and in the case of implantable devices at least 15 years, after the last product has been manufactured, make available to the national authorities:

i. The declaration of conformity.

ii. The documentation referred to in Section 2.

iii. The certificates referred to in Sections 5.2 and 6.4.

iv. Where appropriate, the type-examination certificate referred to in Annex III.

8. Application to devices in Class IIa

In line with Article 11 (2), this Annex may apply to products in Class IIa, subject to the following:

8.1. In derogation from Sections 1 and 2, by virtue of the declaration of conformity the manufacturer ensures and declares that the products in Class IIa are manufactured in conformity with the technical documentation referred to in Section 3 of Annex VII and meet the requirements of this Directive which apply to them;

8.2. In derogation from Sections 1, 2, 5, and 6, the verifications conducted by the notified body are intended to confirm the conformity of the products in Class IIa with the technical documentation referred to in Section 3 of Annex VII.

9. Application to devices referred to in Article 1(4a)

In the case of section 5, upon completing the manufacture of each batch of devices referred to in Article 1(4a), and in the case of verification under section 6, the manufacturer shall inform the notified body of the release of this batch of devices and send to it the official certificate concerning the release of the batch of human blood derivative used in the device issued by a State laboratory or a laboratory designated for that purpose by a Member State in accordance with Article 114(2) of Directive 2001/83/EC.

- **Annex V: EC Declaration of Conformity-Production Quality Assurance:** It is important to understand that while Annex V shares some similarities with Annex II when it comes to QMS requirements. It doesn't include design requirements. This means it requires you to have a QMS for your manufacturing and a final inspection for your devices, but it does not require you to include a design system. This means there may not be a design history file (DHF) in your technical documentation. It should be noted that technical documentation is still a requirement and, in some cases, Annex V will be combined with Annex III (EC Type Examination).

Annex V from the MDD is highlighted below:

Annex V of the MDD 93/42/EEC: EC Declaration of Conformity Production Quality Assurance

1. The manufacturer must ensure application of the quality system approved for the manufacture of the products concerned and carry out the final inspection, as specified in Section 3, and is subject to the Community surveillance referred to in Section 4.

2. The EC declaration of conformity is the part of the procedure whereby the manufacturer who fulfills the obligations imposed by Section 1 ensures and declares that the products concerned conform to the type described in the EC type-examination certificate and meet the provisions of this Directive which apply to them.

The manufacturer must affix the CE marking in accordance with Article 17 and draw up a written declaration of conformity. This declaration must cover one or more medical devices manufactured, clearly identified by means of product name, product code or other unambiguous reference, and must be kept by the manufacturer.

3. Quality system

3.1. The manufacturer must lodge an application for assessment of his quality system with a notified body.

The application must include:

i. The name and address of the manufacturer.

ii. All the relevant information on the product or product category covered by the procedure.

iii. A written declaration that no application has been lodged with any other notified body for the same products.

iv. The documentation on the quality system.

v. An undertaking to fulfill the obligations imposed by the quality system is approved.

vi. An undertaking to maintain the practicability and effectiveness of the approved quality system.

vii. Where appropriate, the technical documentation on the types approved and a copy of the EC type-examination certificates.

viii. An undertaking by the manufacturer to institute and keep up to date a systematic procedure to review experience gained from devices in the post-production phase, including the provisions referred to in Annex X, and to implement appropriate means to apply any necessary corrective action. This undertaking must include an obligation for the manufacturer to notify the competent authorities of the following incidents immediately on learning of them:

a. Any malfunction or deterioration in the characteristics and/or performance of a device, as well as any inadequacy in the labeling or the instructions for use which might lead to or might have led to the death of a patient or user or to a serious deterioration in his state of health;

b. Any technical or medical reason connected with the characteristics or performance of a device for the reasons referred to in subparagraph (i) above leading to a systematic recall of devices of the same type by the manufacturer.

3.2. Application of the quality system must ensure that the products conform to the type described in the EC type-examination certificate.

All the elements, requirements, and provisions adopted by the manufacturer for his quality system must be documented in a systematic and orderly manner in the form of written policy statements and procedures. This quality system documentation must permit uniform interpretation of the quality policy and procedures such as quality programs, plans, manuals, and records.

It must include in particular an adequate description of:

i. The manufacturer's quality objectives;

ii. The organization of the business and in particular:

a. The organizational structures, the responsibilities of the managerial staff and their organizational authority where manufacture of the products is concerned.

b. The methods of monitoring the efficient operation of the quality system and in particular its ability to achieve the desired quality of product, including control of products which fail to conform.

c. Where the manufacture and/or final inspection and testing of the products, or elements thereof, are carried out by a third party, the methods of monitoring the efficient operation of the quality system and in particular the type and extent of control applied to the third party;

iii. The inspection and quality assurance techniques at the manufacturing stage and in particular:

a. The processes and procedures which will be used, particularly as regards sterilization, purchasing, and the relevant documents.

b. The product identification procedures drawn up and kept up to date from drawings, specifications or other relevant documents at every stage of manufacture;

iv. The appropriate tests and trials to be carried out before, during, and after manufacture, the frequency with which they will take place, and the test equipment used; it must be possible adequately to trace back the calibration of the test equipment.

3.3. The notified body must audit the quality system to determine whether it meets the requirements referred to in Section 3.2. It must presume that quality systems that implement the relevant harmonized standards conform to these requirements.

The assessment team must include at least one member with past experience of assessments of the technology concerned. The assessment procedure must include an inspection on the manufacturer's premises and, in duly substantiated cases, on the premises of the manufacturer's suppliers to inspect the manufacturing processes.

The decision must be notified to the manufacturer after the final inspection and contain the conclusions of the inspection and a reasoned assessment.

3.4. The manufacturer must inform the notified body which approved the quality system of any plan for substantial changes to the quality system.

The notified body must assess the changes proposed and verify whether, after these changes, the quality system still meets the requirements referred to in Section 3.2.

After the above-mentioned information has been received, the decision is notified to the manufacturer. It must contain the conclusions of the inspection and a reasoned assessment.

4. Surveillance

4.1. The aim of surveillance is to ensure that the manufacturer duly fulfills the obligations imposed by the approved quality system.

4.2. The manufacturer authorizes the notified body to carry out all the necessary inspections and must supply it with all relevant information, in particular:

i. The documentation on the quality system.

ii. The technical documentation.

iii. The data stipulated in the part of the quality system relating to manufacture, such as inspection reports and test data, calibration data, qualification reports of the personnel concerned, etc.

4.3. The notified body must periodically carry out appropriate inspections and assessments to make sure that the manufacturer applies the approved quality system and supply the manufacturer with an assessment report.

4.4. In addition, the notified body may pay unannounced visits to the manufacturer. At the time of such visits, the notified body may, where necessary, carry out or ask for tests in order to check that the quality system is working properly. It must provide the manufacturer with an inspection report and, if a test has been carried out, with a test report.

5. Administrative provisions

5.1. The manufacturer or his authorized representative must, for a period ending at least five years, and in the case of implantable devices at least 15 years, after the last product has been manufactured, make available to the national authorities:

i. The declaration of conformity.

ii. The documentation referred to in the fourth indent of Section 3.1.

iii. The changes referred to in Section 3.4.

iv. The documentation referred to in the seventh indent of Section 3.1.

v. The decisions and reports from the notified body as referred to in Sections 4.3 and 4.4.

vi. Where appropriate, the type-examination certificate referred to in Annex III.

6. Application to devices in Class IIa

In line with Article 11(2), this Annex may apply to products in Class IIa, subject to the following:

6.1. By way of derogation from Sections 2, 3.1 and 3.2, by virtue of the declaration of conformity the manufacturer ensures and declares that the products in Class IIa are manufactured in conformity with the technical documentation referred to in Section 3 of Annex VII and meet the requirements of this Directive which apply to them.

6.2.	For devices in Class IIa, the notified body shall assess, as part of the assessment in Section 3.3, the technical documentation as described in Section 3 of Annex VII for at least one representative sample for each device subcategory for compliance with the provisions of this Directive.
6.3.	In choosing representative sample(s) the notified body shall take into account the novelty of the technology, similarities in design, technology, manufacturing, and sterilization methods, the intended use and the results of any previous relevant assessments (e.g., with regard to physical, chemical or biological properties) that have been carried out in accordance with this directive. The notified body shall document and keep available to the competent authority its rationale for the sample(s) taken.
6.4.	Further samples shall be assessed by the notified body as part of the surveillance assessment referred to in Section 4.3.
7.	Application to devices referred to in Article 1(4a)
Upon completing the manufacture of each batch of devices referred to in Article 1(4a), the manufacturer shall inform the notified body of the release of the batch of devices and send to it the official certificate concerning the release of the batch of human blood derivative used in the device issued by a State laboratory or a laboratory designated for that purpose by a Member State in accordance with Article 114(2) of Directive 2001/83/ EC.	

- **Annex VI: EC Declaration of Conformity-Product Quality Assurance:** Annex VI requires a quality system for final inspection and testing, but a manufacturer won't need a QMS that includes design or manufacturing. For that reason, this type of assessment is typically combined with Annex III for class IIb devices. This can, however, result in a lengthy review process for the device. This assessment may also not be feasible if the devices require a special type of manufacturing process.

Annex VI from the MDD is highlighted below:

Annex VI of the MDD 93/42/EEC: EC Declaration of Conformity Product Quality Assurance	
1.	The manufacturer must ensure application of the quality system approved for the final inspection and testing of the product, as specified in Section 3 and must be subject to the surveillance referred to in Section 4.
In addition, for products placed on the market in sterile condition, and only for those aspects of the manufacturing process designed to secure and maintain sterility, the manufacturer must apply the provisions of Annex V, Sections 3 and 4.	

2. The EC declaration of conformity is the part of the procedure whereby the manufacturer who fulfills the obligations imposed by Section 1 ensures and declares that the products concerned conform to the type described in the EC type-examination certificate and meet the provisions of this Directive which apply to them.

The manufacturer affixes the CE marking in accordance with Article 17 and draws up a written declaration of conformity. This declaration must cover one or more medical devices manufactured, clearly identified by means of product name, product code or other unambiguous reference, and be kept by the manufacturer. The CE marking must be accompanied by the identification number of the notified body which performs the tasks referred to in this Annex.

3. Quality system

3.1. The manufacturer lodges an application for assessment of his quality system with a notified body.

The application must include:

i. The name and address of the manufacturer.

ii. All the relevant information on the product or product category covered by the procedure.

iii. A written declaration specifying that no application has been lodged with any other notified body for the same products.

iv. The documentation on the quality system.

v. An undertaking by the manufacturer to fulfill the obligations imposed by the quality system approved.

vi. An undertaking by the manufacturer to keep the approved quality system adequate and efficacious.

vii. Where appropriate, the technical documentation on the types approved and a copy of the EC type-examination certificates.

viii. An undertaking by the manufacturer to institute and keep up to date a systematic procedure to review experience gained from devices in the post-production phase, including the provisions referred to in Annex X, and to implement appropriate means to apply any necessary corrective action. This undertaking must include an obligation for the manufacturer to notify the competent authorities of the following incidents immediately on learning of them:

a. Any malfunction or deterioration in the characteristics and/or performance of a device, as well as any inadequacy in the labeling or the instructions for use which might lead to or might have led to the death of a patient or user or to a serious deterioration in his state of health;

b. Any technical or medical reason connected with the characteristics or the performance of a device for the reasons referred to in subparagraph

c. Leading to a systematic recall of devices of the same type by the manufacturer.

3.2. Under the quality system, each product or a representative sample of each batch is examined and the appropriate tests defined in the relevant standard(s) referred to in Article 5 or equivalent tests are carried out to ensure that the products conform to the type described in the EC type-examination certificate and fulfill the provisions of this Directive which apply to them. All the elements, requirements, and provisions adopted by the manufacturer must be documented in a systematic and orderly manner in the form of written measures, procedures, and instructions. This quality system documentation must permit uniform interpretation of the quality programs, quality plans, quality manuals and quality records.

It must include in particular an adequate description of:

i. The quality objectives and the organizational structure, responsibilities, and powers of the managerial staff with regard to product quality.

ii. The examinations and tests that will be carried out after manufacture; it must be possible to trace back the calibration of the test equipment adequately.

iii. The methods of monitoring the efficient operation of the quality system.

iv. The quality records, such as reports concerning inspections, tests, calibration, and the qualifications of the staff concerned, etc.

v. Where the final inspection and testing of the products, or elements thereof, are carried out by a third party, the methods of monitoring the efficient operation of the quality system and in particular the type and extent of control applied to the third party.

The aforementioned checks do not apply to those aspects of the manufacturing process designed to secure sterility.

3.3. The notified body audits the quality system to determine whether it meets the requirements referred to in Section 3.2. It must presume that quality systems that implement the relevant harmonized standards conform to these requirements.

The assessment team must include at least one member with past experience of assessments of the technology concerned. The assessment procedure must include an inspection on the manufacturer's premises and, in duly substantiated cases, on the premises of the manufacturer's suppliers to inspect the manufacturing processes.

The decision must be notified to the manufacturer. It must contain the conclusions of the inspection and a reasoned assessment.

3.4. The manufacturer must inform the notified body which approved the quality system of any plan for substantial changes to the quality system.

The notified body must assess the changes proposed and verify whether after these changes the quality system will still meet the requirements referred to in Section 3.2.

After receiving the above-mentioned information, it must notify the manufacturer of its decision. This decision must contain the conclusions of the inspection and a reasoned assessment.

4. Surveillance

4.1. The aim of surveillance is to ensure that the manufacturer duly fulfills the obligations imposed by the approved quality system.

4.2. The manufacturer must allow the notified body access for inspection purposes to the inspection, testing, and storage locations and supply it with all relevant information, in particular:

i. The documentation on the quality system.

ii. the technical documentation.

iii. the quality records, such as inspection reports, test data, calibration data, qualification reports of the staff concerned, etc.

4.3. The notified body must periodically carry out appropriate inspections and assessments to make sure that the manufacturer applies the quality system and must supply the manufacturer with an assessment report.

4.4. In addition, the notified body may pay unannounced visits to the manufacturer. At the time of such visits, the notified body may, where necessary, carry out or ask for tests in order to check that the quality system is working properly and that the production conforms to the requirements of the Directive which apply to it. To this end, an adequate sample of the final products, taken on-site by the notified body, must be examined and the appropriate tests defined in the relevant standard(s) referred to in Article 5 or equivalent tests must be carried out. Where one or more of the samples fails to conform, the notified body must take the appropriate measures.

It must provide the manufacturer with an inspection report and, if a test has been carried out, with a test report.

5. Administrative provisions

5.1. The manufacturer or his authorized representative must, for a period ending at least five years, and in the case of implantable devices at least 15 years, after the last product has been manufactured, make available to the national authorities:

i. the declaration of conformity.

ii. the documentation referred to in the seventh indent of Section 3.1.

iii. the changes referred to in Section 3.4.

iv. the decisions and reports from the notified body as referred to in the final indent of Section 3.4 and in Sections 4.3 and 4.4.

v. where appropriate, the certificate of conformity referred to in Annex III.

6. Application to devices in Class IIa

In line with Article 11(2), this Annex may apply to products in Class IIa, subject to the following:

6.1. By way of derogation from Sections 2, 3.1 and 3.2, by virtue of the declaration of conformity the manufacturer ensures and declares that the products in Class IIa are manufactured in conformity with the technical documentation referred to in Section 3 of Annex VII and meet the requirements of this Directive which apply to them.

6.2. For devices in Class IIa, the notified body shall assess, as part of the assessment in Section 3.3, the technical documentation as described in Section 3 of Annex VII for at least one representative sample for each device subcategory for compliance with the provisions of this Directive.

6.3.	In choosing representative sample(s) the notified body shall take into account the novelty of the technology, similarities in design, technology, manufacturing, and sterilization methods, the intended use and the results of any previous relevant assessments (e.g., with regard to physical, chemical or biological properties) that have been carried out in accordance with this Directive. The notified body shall document and keep available to the competent authority its rationale for the sample(s) taken.
6.4.	Further samples shall be assessed by the notified body as part of the surveillance assessment referred to in Section 4.3.

- **Annex VII: EC Declaration of Conformity:** Annex VII requires you to have a systematic procedure in place to collect post-market information about the devices in use, when in the field/on the market. During the assessment, you will be required to provide technical documentation demonstrate conformity with Essential Requirements.

Annex VII from the MDD is highlighted below:

Annex VII of the MDD 93/42/EEC: EC Declaration of Conformity	
1.	The EC declaration of conformity is the procedure whereby the manufacturer or his authorized representative who fulfills the obligations imposed by Section 2 and, in the case of products placed on the market in a sterile condition and devices with a measuring function, the obligations imposed by Section 5 ensures and declares that the products concerned meet the provisions of this Directive which apply to them.
2.	The manufacturer must prepare the technical documentation described in Section 3. The manufacturer or his authorized representative must make this documentation, including the declaration of conformity, available to the national authorities for inspection purposes for a period ending at least five years after the last product has been manufactured. In the case of implantable devices, the period shall be at least 15 years after the last product has been manufactured.
3.	The technical documentation must allow assessment of the conformity of the product with the requirements of the Directive. It must include in particular:
i.	a general description of the product, including any variants planned and its intended use(s).

ii. design drawings, methods of manufacture envisaged and diagrams of components, sub-assemblies, circuits, etc.

iii. the descriptions and explanations necessary to understand the above-mentioned drawings and diagrams and the operations of the product.

iv. the results of the risk analysis and a list of the standards referred to in Article 5, applied in full or in part, and descriptions of the solutions adopted to meet the essential requirements of the Directive if the standards referred to in Article 5 have not been applied in full.

v. in the case of products placed on the market in a sterile condition, description of the methods used and the validation report.

vi. the results of the design calculations and of the inspections carried out, etc.; if the device is to be connected to other device(s) in order to operate as intended, proof must be provided that it conforms to the essential requirements when connected to any such device(s) having the characteristics specified by the manufacturer.

vii. the solutions adopted as referred to in Annex I, Chapter I, Section 2.

viii. the preclinical evaluation.

ix. the clinical evaluation in accordance with Annex X.

x. the label and instructions for use.

4. The manufacturer shall institute and keep up to date a systematic procedure to review experience gained from devices in the post-production phase, including the provisions referred to in Annex X, and to implement appropriate means to apply any necessary corrective actions, taking account of the nature and risks in relation to the product. He shall notify the competent authorities of the following incidents immediately on learning of them:

i. any malfunction or deterioration in the characteristics and/or performance of a device, as well as any inadequacy in the labeling or the instructions for use which might lead to or might have led to the death of a patient or user or to a serious deterioration in his state of health;

ii. any technical or medical reason connected with the characteristics on the performance of a device for the reasons referred to in subparagraph (i) leading to systematic recall of devices of the same type by the manufacturer.

5. With products placed on the market in sterile condition and Class I devices with a measuring function, the manufacturer must observe not only the provisions laid down in this Annex but also one of the procedures referred to in Annex II, IV, V or VI. Application of the above-mentioned Annexes and the intervention by the notified body is limited to:

i. in the case of products placed on the market in sterile condition, only the aspects of manufacture concerned with securing and maintaining sterile conditions.

ii. in the case of devices with a measuring function, only the aspects of manufacture concerned with the conformity of the products with the metrological requirements. Section 6.1. of this Annex is applicable.

6. Application to devices in Class IIa

In line with Article 11 (2), this Annex may apply to products in Class IIa, subject to the following derogation:

6.1. where this Annex is applied in conjunction with the procedure referred to in Annex IV, V or VI, the declaration of conformity referred to in the above-mentioned Annexes forms a single declaration. As regards the declaration based on this Annex, the manufacturer must ensure and declare that the product design meets the provisions of this Directive which apply to it.

A manufacturer must remember that the classification of the device dictates the appropriate conformity assessment procedure that should be used. These are further explained below as designated by the MDD (Table 4.1).

Table 4.1: Overview of Device Classification and Appropriate Conformity Assessment Procedures

Conformity Assessment Procedures	**Classes**					
Annexes	**I**	**I (Sterile)**	**I (Measuring)**	**IIA**	**IIB**	**III**
II (+sect 4)	–	–	–	–	–	X
II (–sect 4)	–	X	X	X	X	–
III	–	–	–	–	X	X
IV	–	X	X	X	X	X
V	–	X	X	X	X	X
VI	–	X	X	X	X	
VII	X	X	X	X	–	–

4.2.1 Class I Medical Device Conformity Assessment Routes

For a class I medical device, the manufacturer is responsible for ensuring that his product complies with all the relevant Essential Requirements of the Directive. It is the manufacturer's responsibility to draw up a written statement indicating this, also known as a self-declaration. Class I medical devices without a measuring function and which are supplied in a non-sterile condition does not require the involvement of a Notified Body. It should also be noted that conformity to the International and European ISO 13485 standard is voluntary.

Manufacturers of sterile products and devices with a measuring function must apply to a Notified Body for certification of those manufacturing aspects related to sterility or metrology.

Once the manufacturer is satisfied that his product meets all the relevant Essential Requirements, the manufacturer, or his European Authorized Representative can register with the competent authority (CA), followed by affixing a CE mark on the product. The device is then ready to be placed on the European market (Figure 4.1).

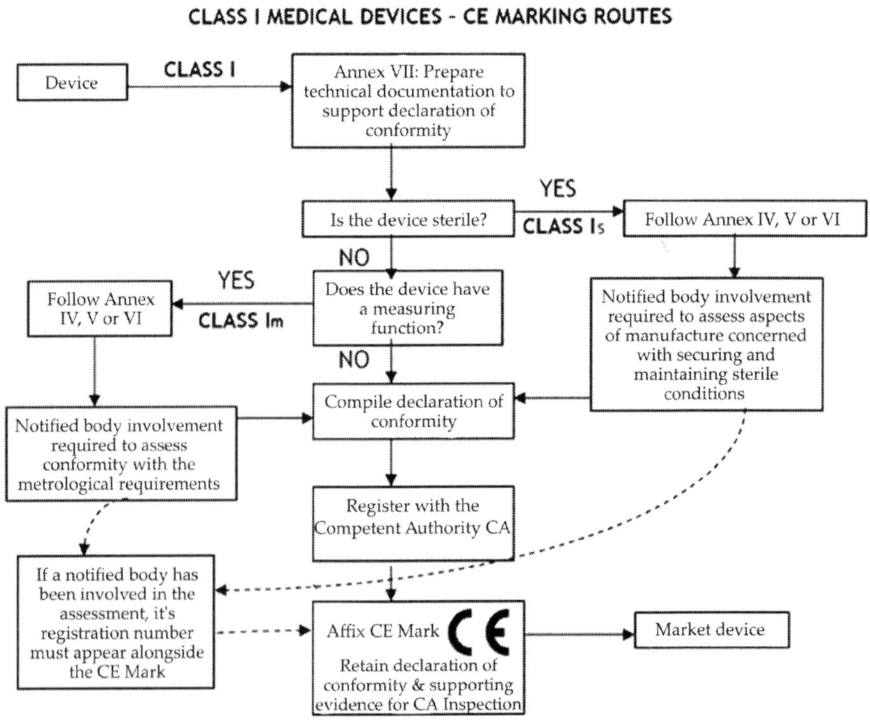

Figure 4.1: Class I medical devices CE marking routes.

Source: http://www.ce-marking.com/medical-devices-class-i.html.

Simplified steps for class I medical device compliance:	
1.	Classification: ensure the device is a Class I medical device
2.	Choose relevant Conformity Assessment Route
3.	Compile the Technical File
4.	Draw up the Declaration of Conformity
5.	Appoint an Authorized Representative
6.	Register with the Competent Authority
7.	Affix CE marking and market the products
8.	Maintain vigilance monitoring and Post Market Surveillance

4.2.2. Class IIa Medical Device Conformity Assessment Routes

For class IIa medical devices (other than devices which are custom-made or intended for clinical investigations), the manufacturer shall either:

1. Follow the procedure relating to the EC declaration of conformity (DoC) set out in Annex II (full quality assurance); in this case, point 4 of Annex II is not applicable; or
2. Follow the procedure relating to the EC declaration of conformity set out in Annex VII, coupled with either:
   - i. the procedure relating to the EC verification set out in Annex IV;
   - ii. the procedure relating to the EC declaration of conformity set out in Annex V (production quality assurance); or
   - iii. the procedure relating to the EC declaration of conformity set out in Annex VI (product quality assurance).
3. For Class IIa products, the declaration must be backed up in all cases with conformity assessment by Notified Body and this assessment may consist of:
   1. Examination and testing of each product or homogenous batch of products (Annex IV);
   2. Audit of the production quality assurance system (Annex V:) ISO 13485 (excluding Design);
   3. Audit of final inspection and testing (Annex VI:) ISO 13485 (excluding Design and Manufacture); or
   4. Audit of the full quality assurance system (Annex II) ISO 13485.

This is decided upon by the manufacturer. Once the manufacturer has received certification from the Notified Body they may CE mark the product and place them on the market (Figure 4.2).

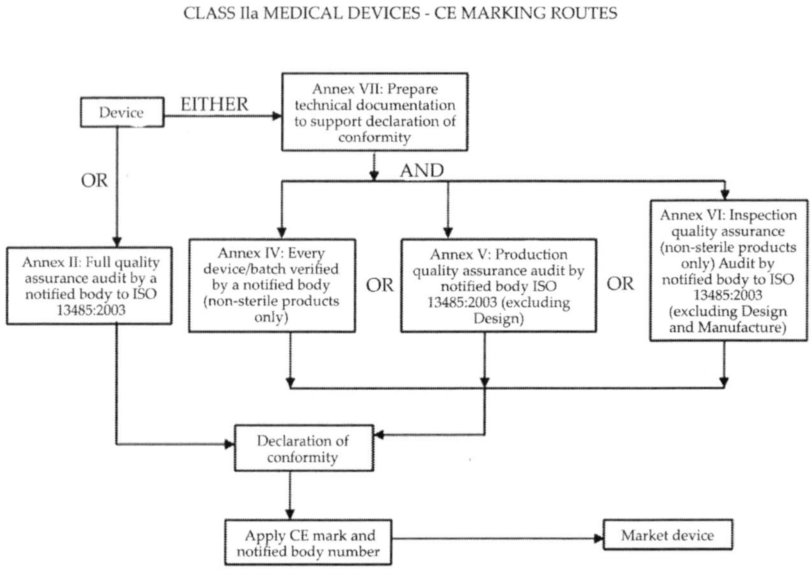

Figure 4.2: Class IIa medical devices CE marking routes.

Simplified steps for class IIa medical device compliance:	
1.	Classification: ensure the device is a Class IIa medical device
2.	Choose Conformity Assessment Route
3.	Compile the Technical File
4.	Obtain certification from a Notified Body
5.	Draw up the Declaration of Conformity
6.	Appoint an Authorized Representative
7.	Hold the Tech Files for inspection by the Competent Authority
8.	Affix CE marking and market the products
9.	Maintain vigilance monitoring and Post Market Surveillance

4.2.3. Class IIb Medical Device Conformity Assessment Routes

In the case of devices falling within Class IIb, (other than devices which are custom-made or intended for clinical investigations), the manufacturer shall, either:

- follow the procedure relating to the EC declaration of conformity set out in Annex II (full quality assurance); in this case, point 4 of Annex II is not applicable;
- follow the procedure relating to the EC type-examination set out in Annex III, coupled with:

i. the procedure relating to the EC verification set out in Annex IV;

ii. the procedure relating to the EC declaration of conformity set out in Annex V (production quality assurance); or
iii. the procedure relating to the EC declaration of conformity set out in Annex VI (product quality assurance).

There are two routes:

1. a Notified Body must carry out either an Annex II audit of the full quality assurance system (ISO 13485), or
2. a type-examination (Annex III) plus one of the three options given here:
    i. Examination and testing of each product or homogenous batch of products (Annex IV);
    ii. Audit of the production quality assurance system (Annex V:) ISO 13485 (excluding Design); or
    iii. Audit of final inspection and testing (Annex VI:) ISO 13485 (excluding Design and Manufacture).

Once the manufacturer has received certification from the Notified Body, they may CE mark the product and place them on the market (Figure 4.3).

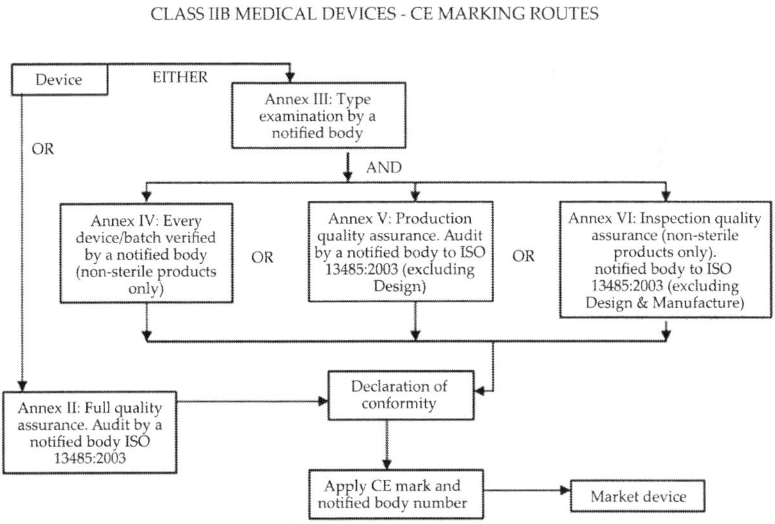

Figure 4.3: Class IIb medical devices CE marking routes.

Simplified steps for class IIb medical device compliance:	
1.	Classification: ensure the device is a Class IIb medical device
2.	Choose Conformity Assessment Route
3.	Compile the Technical File
4.	Obtain certification from a Notified Body
5.	Draw up the Declaration of Conformity
6.	Appoint an Authorized Representative
7.	Hold the Tech Files for inspection by the Competent Authority
8.	Affix CE marking and market the products
9.	Maintain vigilance monitoring and Post Market Surveillance

4.2.4. Class III Medical Device Conformity Assessment Routes

This route to conformity is similar to those for Class IIb devices but additionally requires the manufacturer to submit the design dossier to the NB for approval under audit of the full quality assurance system (Annex II) and do not allow the Annex VI option. Active implantable medical devices (AIMD) and *in vitro* diagnostic (IVD) medical devices are also subject to conformity procedures. In the case of devices falling within Class III (other than devices which are custom-made or intended for clinical investigations) the manufacturer shall either:

- follow the procedure relating to the EC declaration of conformity set out in Annex II (full quality assurance);
- follow the procedure relating to the EC type-examination set out in Annex III, coupled with:
    i. the procedure relating to the EC verification set out in Annex IV; or
    ii. the procedure relating to the EC declaration of conformity set out in Annex V (production quality assurance).

Class III controls are similar to those for Class IIb devices but additionally require the manufacturer to submit the design dossier to the Notified Body for approval under Annex II and do not allow the Annex III/Annex VI option.

There are two routes:

1. a Notified Body must carry out either an Annex II audit of the full quality assurance system (ISO 13485), plus that the manufacturer must submit the design dossier to the Notified Body for approval under Annex II, or
2. a type-examination (Annex III) plus one of the two options given here:

i. Examination and testing of each product or homogenous batch of products (Annex IV); or
ii. Audit of the production quality assurance system (Annex V:) ISO 13485 (excluding Design)

Once the manufacturer has received certification from the Notified Body they may CE mark the product and place them on the market (Figure 4.4).

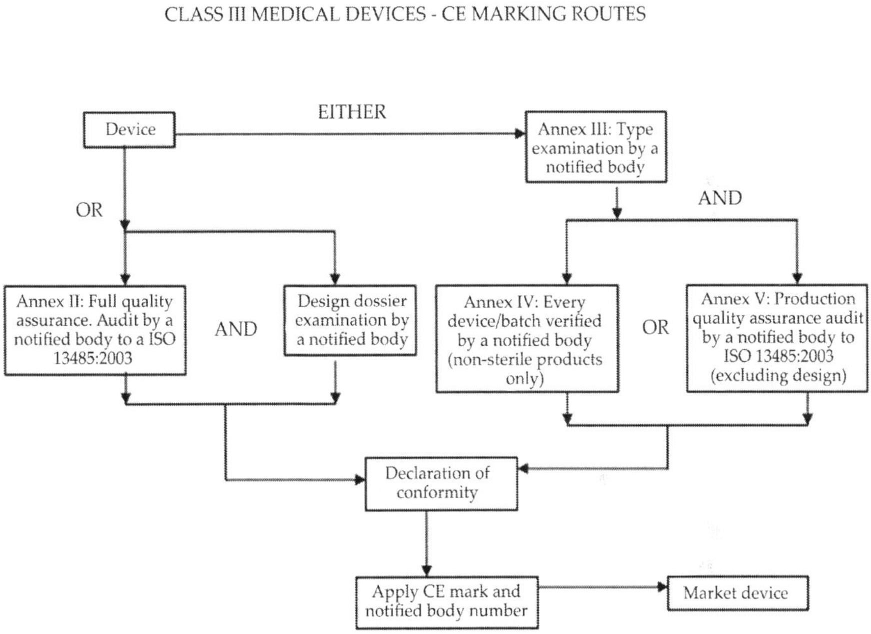

Figure 4.4: Class III medical devices CE marking routes.

Simplified steps for class IIb medical device compliance:	
1.	Classification: ensure the device is a Class III medical device
2.	Choose Conformity Assessment Route
3.	Compile the Technical File
4.	Obtain certification from a Notified Body
5.	Draw up the Declaration of Conformity
6.	Appoint an Authorized Representative
7.	Hold the Tech Files for inspection by the Competent Authority
8.	Affix CE marking and market the products
9.	Maintain vigilance monitoring and Post Market Surveillance

4.2.5. IVD-In Vitro Diagnostic Medical Device Conformity Assessment Routes

1. For all devices referred to in List A in Annex II other than those intended for performance evaluation, the manufacturer shall, in order to affix the CE marking either:
   i. follows the procedure relating to the EC declaration of conformity set out in Annex IV (full quality assurance); or
   ii. follow the procedure relating to EC type-examination set out in Annex V coupled with the procedure relating to the EC declaration of conformity set out in Annex VII (production quality assurance).
2. For all devices referred to in List B in Annex II other than those intended for performance evaluation, the manufacturer shall for the purposes of affixing the CE marking, follow either:
   i. the procedure relating to the EC declaration of conformity set out in Annex IV (full quality assurance); or
   ii. the procedure relating to EC type-examination set out in Annex V couplet with:
      a. the procedure relating to EC verification set out in Annex VI, or
      b. the procedure relating to the EC declaration of conformity set out in Annex VII (production quality assurance).
3. Other than those covered by Annex II:
   i. **Self-Testing:** For all devices for self-testing other than those covered by Annex II and devices for performance evaluation, the manufacturer shall, prior to the drawing up the EC declaration of conformity in Annex III, fulfill the supplementary requirements set out in Annex III, point 6. Instead of applying this procedure, the manufacturer may follow the procedure referred to for products in List A or List B in Annex II.
   ii. **Other/General:** For all devices other than those covered by Annex II and devices for performance evaluation, the manufacturer shall, in order to affix the CE marking, follow the procedure referred to in Annex III and draw up the EC declaration of conformity required before placing the devices on the market.

4. In the case of devices for performance evaluation, the manufacturer shall follow the procedure referred to in Annex VIII and draw up the statement set out in that Annex before such devices are made available.

This provision does not affect national regulations relating to the ethical aspects of carrying out performance evaluation studies using tissues or substances of human origin.

4.3. UNDERSTANDING THE CHANGES IN CONFORMITY ASSESSMENTS UNDER THE MDR

Under the MDR, conformity assessment procedures remain essentially the same. Manufacturers still need to conduct a conformity assessment before placing their devices on the market. Similarly, for higher risk devices, the involvement of a notified body is also required. There is also an added requirement for the notified bodies (NB) to consult with health authorities and EU expert panels, and this is expected to prolong quite remarkably the conformity assessment route for class III devices. Apart from Annex numbers being changed, the approach for conducting the conformity assessment routes remains essentially similar to that of the MDD. The number of general safety and performance requirements (GSPRs) have also increased under the MDR in comparison to the Essential Requirements under the MDD, but this will be discussed further later on in this volume.

Device classification will determine the conformity assessment route chosen under the MDR. Classification of a device is regulated by Annex VIII of the MDR, and the device itself can be grouped into one of four classes: I, IIa, IIb, and III; taking into account the intended purpose and related risks. Depending on the risk class chosen, the manufacturer may choose the appropriate conformity assessment route to demonstrate compliance with the regulation. Three levels of control are presented and this is dependent on the perceived risk associated with the device:

- "Self-certification" where no external party is involved in the conformity assessment;
- Certification by a NB; and
- The involvement of a CA or expert panels through a commission or medical device coordination group (MDCG).

In the MDR, article 52 outlines the conformity assessment procedures as well as the respective annexes which have essentially been renumbered and renamed between Annex IX to XI. Some of the major changes to conformity assessment routes between the MDD and MDR are listed below:

- A new class for reusable surgical instruments (Class Ir) has been introduced to control the safety around reusable characteristics of surgical instruments. This class also requires a notified body to assess the QMS.
- The MDR introduces new classification rules which may essentially lead to a higher classification of devices than that defined by the MDD. This means there will be an increase in the amount of clinical data required as well as more rigorous post-market surveillance.
- The full quality assurance route is replaced by the conformity assessment based on quality system assurance and assessment of the technical documentation in Annex IX.
- Conformity assessment procedures will generally be more demanding on a manufacturer and also more likely to take longer, especially in those cases where a notified body has to consult with health authorities and EU expert panels for the assessment of high-risk devices.

4.4. UNDERSTANDING THE CONFORMITY ASSESSMENT ROUTES IN THE MDR

4.4.1. Annex IX (QMS and Technical Documentation)

Annex IX is used when a full QMS is implemented by the manufacturer. With this conformity assessment route, a review of technical documentation is needed with or without the issuance of a certificate. Depending on the device classification, full or partial compliance with the Annex IX is required.

Annex IX comprises of three parts: (1) Assessment of QMS via NB audit with issuance of CE Marking certificate; (2) Assessment of Technical File/ Design Dossier via NB audit with issuance of CE Marking certificate; (3) General provisions for record retention and availability of documentation to competent authority (CA).

4.4.2. Annex X (Type-Examination)

Annex X is used when a manufacturer wants to certify a device based on a representative sample. The NB examines and/or tests the representative sample and associated technical documentation to determine if the device meets MDR requirements with a focus on the GSPR.

4.4.3. Annex XI (Product Conformity Verification)

Annex XI is generally used in association with Annex X or in combination with technical documentation annexes (Annexes II and III) for low-risk devices. This annex is composed of two parts: (1) Production Quality Assurance via NB audit with issuance of CE marking certificate for ability to produce and test a device; (2) Product Verification via NB audit with issuance of CE marking certificate supporting the conformity of a specific batch of devices.

A technical file (or technical documentation) is required to show compliance with Annex II (technical documentation) and Annex III (technical documentation on Post-Market Surveillance-PMS) but this will depend on the device and route of compliance (Table 4.2).

Table 4.2: Comparison of Conformity Assessment Routes between the MDD and MDR

MDD	MDR	Focus of Annex
Annex II excluding section 4 Full quality assurance	Annex IX chapters I and III Quality management system	QMS based Design, manufacture, final inspection
Annex II section 4 Design examination	Annex IX Chapter II Technical documentation	Product-based; Type-testing and documentation review
Annex IV verification	Annex IX-Part B Product verification	Product-based Individual devices tested
Annex V Production quality assurance	Annex XI-Part A Production quality assurance	Q<S based, manufacture, final inspection
Annex VI product quality assurance	No equivalent	QMS based, Final inspection
Annex VII Declaration of conformity	Article 19 + annex II, III	For class I devices

4.4.4. Class I Medical Device Conformity Assessment Route Under MDR

Under the MDR, the only conformity assessment route for a self-certified Class I medical device is to maintain technical documentation in compliance with Annex II and III. For Class I devices supplied sterile (Is), with a measuring function (Im), or those that are reusable surgical instruments (Ir), a limited QMS must be in place to control the production (as per Annex XI Part A of the MDR) or to control the special characteristic features (e.g., sterility, measuring, reusability) (Annex IX-Chapter I). For class Is, Im, and Ir devices, an NB will be involved to control how the QMS manages those specific features in relation to the conformity assessment procedure selected (Figure 4.5).

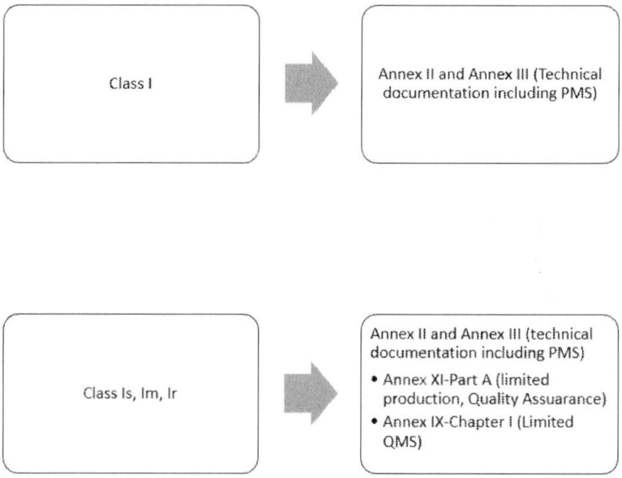

Figure 4.5: Conformity assessment procedures for class I devices under MDR 2017/745.

4.4.5. Class IIa Medical Device Conformity Assessment Route Under MDR

Under the MDR, the conformity assessment procedure for a Class IIa device through the review of full QMS (Annex IX) is identical to the procedure for a Class IIb non-active and non-implantable device. Alternatively, a manufacturer may build technical documentation aligned with Annex II and III and select a route of conformity assessment based on the production control (Annex XI) (Figure 4.6).

class IIa

- Annex IX-chapter I & section 4: full QMS and Technical Documentation for representative device without expert review
- Annex II & Annex III: Technical documentation (including PMS) review for representative device
  - Annex XI-section 10: production quallity assurance
  - Annex XI-section 18: product verification

Figure 4.6: Conformity assessment procedures for class IIa devices under MDR 2017/745.

4.4.6. Class IIb Medical Device Conformity Assessment Route Under MDR

The level of control for Class IIb devices is similar to Class III devices, though there is no separate CE certificate issued for the technical documentation assessment. In addition, involvement from an expert panel is not required in the review process unless the Class IIb device is active and intended to remove or administer a medicinal substance. The technical documentation review is performed on a representative sample of device type with the exception of Class IIb implantable devices, for which 100% review is required (Figure 4.7).

class IIb active with drug administration or removal

- Annex IX-chapter I & section 4: full QMS and Technical documentation for every device with expert review
- Annex X: assessment of type-examination
  - Annex XI-part A: production quality assurance
  - Annex XI-part B: product verification

class IIb

- Annex IX-chapter I & section 4: for class IIb implantable devices*: full QMS and technical documentation for every device without expert review/for class IIb non-implantable devices*: full QMS and technical documentation for representative device without expert review
- Annex X: assessment of type examination
  - Annex XI-part A: product quality assurance
  - Annex XI-part B: product verification

Figure 4.7: Conformity assessment procedures for class IIb devices under MDR 2017/745.

*Note: *Except sutures, staples, dental fillings, dental braces, tooth crowns, screws, wedges, plates, wires, pins, clips, and connectors*

4.4.7. Class III Medical Device Conformity Assessment Route under MDR

Conformity assessment for Class III medical devices is covered by Annex IX and includes a full Quality Assurance audit and full Technical Documentation review. A certificate is issued following the NB QMS audit and a second certificate is issued for every device after review of the technical documentation. An expert panel is also involved in the evaluation for Class III implantable devices.

While standards are voluntary, one way of presuming conformity to the GSPR and meeting the provisions of full quality assurance is to possess harmonized EN ISO 13485 standard certifications, which pertains to the state of the art and expected requirements for a company's QMS in the EU. It should be noted that manufacturers of Class III medical devices also have the option to pursue Annex X, Type-Examination, in combination with Annex XI Part A or Part B and, therefore, with a QMS focused on production and controls (Figure 4.8).

class III

- Annex IX: assessment of full QMS and technical documentation for every device with expert review
- Annex X: assessment of type examination
  - Annex XI-part A: production quality assurance
  - Annex IX-part B: product verification

Figure 4.8: Conformity assessment procedures for class III devices under MDR 2017/745.

4.5. SAFETY AND PERFORMANCE REQUIREMENTS OF A MEDICAL DEVICE

This Essential Requirements section of the MDD focuses on the minimum essential requirements for the design and manufacture of medical devices. It is aimed at ensuring the protection of the health and safety of patients, users, and third parties. For this reason, these requirements stipulate that the principles of safety should be integral to the design of the product and that the product should be suitable for its intended purpose. The essential requirements are defined in Annex I of the MDD. Annex 1 of the Directive lays down a series of essential requirements for the design, safety, and functionality of medical devices. The Essential Requirements of the medical device directive (MDD) apply to all medical devices and cover the following:

- a general requirement for safe design;
- the minimization of risks from contamination;
- compatibility with materials with which they are likely to come into contact;
- minimization of hazards of infection and microbial contamination;
- provision of sufficient accuracy (for devices with a measuring function);
- protection against radiation;
- adequate product marking;
- adequate user instructions.

The specific details and requirements of these essential requirements are laid out in the directive and are further defined below:

Excerpt from the MDD: Article 3: Essential Requirements

Article 3 of the MDD 93/42/EEC: Essential Requirements
The devices must meet the essential requirements set out in Annex I which apply to them, taking account of the intended purpose of the devices concerned. Where a relevant hazard exists, devices which are also machinery within the meaning of Article 2(a) of Directive 2006/42/EC of the European Parliament and of the Council of 17 May 2006 on machinery (5) shall also meet the essential health and safety requirements set out in Annex I to that Directive to the extent to which those essential health and safety requirements are more specific than the essential requirements set out in Annex I to this Directive.

Now let's look at the requirements of Annex I (Essential Requirements) in greater detail.

Annex I of the MDD 93/42/EEC: Essential Requirements

I. General Requirements

1. The devices must be designed and manufactured in such a way that, when used under the conditions and for the purposes intended, they will not compromise the clinical condition or the safety of patients, or the safety and health of users or, where applicable, other persons, provided that any risks which may be associated with their intended use constitute acceptable risks when weighed against the benefits to the patient and are compatible with a high level of protection of health and safety.

This shall include:

i. reducing, as far as possible, the risk of use error due to the ergonomic features of the device and the environment in which the device is intended to be used (design for patient safety), and

ii. consideration of the technical knowledge, experience, education, and training and, where applicable, the medical and physical conditions of intended users (design for lay, professional, disabled or other users).

2. The solutions adopted by the manufacturer for the design and construction of the devices must conform to safety principles, taking account of the generally acknowledged state of the art.

In selecting the most appropriate solutions, the manufacturer must apply the following principles in the following order:

i. eliminate or reduce risks as far as possible (inherently safe design and construction).

ii. where appropriate, take adequate protection measures, including alarms if necessary, in relation to risks that cannot be eliminated.

iii. inform users of the residual risks due to any shortcomings of the protection measures adopted.

3. The devices must achieve the performances intended by the manufacturer and be designed, manufactured, and packaged in such a way that they are suitable for one or more of the functions referred to in Article 1 (2) (a), as specified by the manufacturer.

4. The characteristics and performances referred to in Sections 1, 2 and 3 must not be adversely affected to such a degree that the clinical conditions and safety of the patients and, where applicable, of other persons are compromised during the lifetime of the device as indicated by the manufacturer, when the device is subjected to the stresses which can occur during normal conditions of use.

5. The devices must be designed, manufactured, and packed in such a way that their characteristics and performances during their intended use will not be adversely affected during transport and storage, taking account of the instructions and information provided by the manufacturer.

6. Any undesirable side-effect must constitute an acceptable risk when weighed against the performances intended.

6a. Demonstration of conformity with the essential requirements must include a clinical evaluation in accordance with Annex X.

II. requirements regarding design and construction

7. Chemical, physical, and biological properties

7.1. The devices must be designed and manufactured in such a way as to guarantee the characteristics and performances referred to in Section I on the 'General requirements. Particular attention must be paid to:

i. the choice of materials used, particularly as regards toxicity and, where appropriate, flammability.

ii. the compatibility between the materials used and biological tissues, cells, and body fluids, taking account of the intended purpose of the device.

iii. where appropriate, the results of biophysical or modeling research whose validity has been demonstrated beforehand.

7.2. The devices must be designed, manufactured, and packed in such a way as to minimize the risk posed by contaminants and residues to the persons involved in the transport, storage, and use of the devices and to the patients, taking account of the intended purpose of the product. Particular attention must be paid to the tissues exposed and to the duration and frequency of exposure.

7.3. The devices must be designed and manufactured in such a way that they can be used safely with the materials, substances, and gases with which they enter into contact during their normal use or during routine procedures; if the devices are intended to administer medicinal products they must be designed and manufactured in such a way as to be compatible with the medicinal products concerned according to the provisions and restrictions governing these products and that their performance is maintained in accordance with the intended use.

7.4. Where a device incorporates, as an integral part, a substance which, if used separately, may be considered to be a medicinal product as defined in Article 1 of Directive 2001/83/EC and which is liable to act upon the body with action ancillary to that of the device, the quality, safety, and usefulness of the substance must be verified by analogy with the methods specified in Annex I to Directive 2001/83/ EC.

For the substances referred to in the first paragraph, the notified body shall, having verified the usefulness of the substance as part of the medical device and taking account of the intended purpose of the device, seek a scientific opinion from one of the competent authorities designated by the Member States or the European Medicines Agency (EMEA) acting particularly through its committee in accordance with Regulation (EC) No 726/2004 (1) on the quality and safety of the substance including the clinical benefit/risk profile of the incorporation of the substance into the device. When issuing its opinion, the competent authority or the EMEA shall take into account the manufacturing process and the data related to the usefulness of incorporation of the substance into the device as determined by the notified body. Where a device incorporates, as an integral part, a human blood derivative, the notified body shall, having verified the usefulness of the substance as part of the medical device and taking into account the intended purpose of the device, seek a scientific opinion from the EMEA, acting particularly through its committee, on the quality and safety of the substance including the clinical benefit/risk profile of the incorporation of the human blood derivative into the device.

When issuing its opinion, the EMEA shall take into account the manufacturing process and the data related to the usefulness of incorporation of the substance into the device as determined by the notified body.

Where changes are made to an ancillary substance incorporated in a device, in particular related to its manufacturing process, the notified body shall be informed of the changes and shall consult the relevant medicines competent authority (i.e., the one involved in the initial consultation), in order to confirm that the quality and safety of the ancillary substance are maintained. The competent authority shall take into account the data related to the usefulness of incorporation of the substance into the device as determined by the notified body, in order to ensure that the changes have no negative impact on the established benefit/risk profile of the addition of the substance in the medical device.

When the relevant medicines competent authority (i.e., the one involved in the initial consultation) has obtained information on the ancillary substance, which could have an impact on the established benefit/risk profile of the addition of the substance in the medical device, it shall provide the notified body with advice, whether this information has an impact on the established benefit/risk profile of the addition of the substance in the medical device or not. The notified body shall take the updated scientific opinion into account in reconsidering its assessment of the conformity assessment procedure.

7.5. The devices must be designed and manufactured in such a way as to reduce to a minimum the risks posed by substances leaking from the device. Special attention shall be given to substances which are carcinogenic, mutagenic or toxic to reproduction, in accordance with Annex I to Council Directive 67/548/EEC of 27 June 1967 on the approximation of laws, regulations, and administrative provisions relating to the classification, packaging, and labeling of dangerous substances (1).

If parts of a device (or a device itself) intended to administer and/or remove medicines, body-liquids or other substances to or from the body, or devices intended for transport and storage of such body fluids or substances, contain phthalates which are classified as carcinogenic, mutagenic or toxic to reproduction, of category 1 or 2, in accordance with Annex I to Directive 67/548/EEC, these devices must be labeled on the device itself and/or on the packaging for each unit or, where appropriate, on the sales packaging as a device containing phthalates.

If the intended use of such devices includes treatment of children or treatment of pregnant or nursing women, the manufacturer must provide a specific justification for the use of these substances with regard to compliance with the essential requirements, in particular of this paragraph, within the technical documentation and, within the instructions for use, information on residual risks for these patient groups and, if applicable, on appropriate precautionary measures.

7.6. Devices must be designed and manufactured in such a way as to reduce, as much as possible, risks posed by the unintentional ingress of substances into the device, taking into account the device and the nature of the environment in which it is intended to be used.

8. Infection and microbial contamination

8.1. The devices and manufacturing processes must be designed in such a way as to eliminate or reduce as far as possible the risk of infection to the patient, user, and third parties. The design must allow easy handling and, where necessary, minimize contamination of the device by the patient or vice versa during use.

8.2. Tissues of animal origin must originate from animals that have been subjected to veterinary controls and surveillance adapted to the intended use of the tissues.

Notified bodies shall retain information on the geographical origin of the animals.

Processing, preservation, testing, and handling of tissues, cells, and substances of animal origin must be carried out so as to provide optimal security. In particular safety with regard to viruses and other transmissible agents must be addressed by implementation of validated methods of elimination or viral inactivation in the course of the manufacturing process.

8.3. Devices delivered in a sterile state must be designed, manufactured, and packed in a non-reusable pack and/or according to appropriate procedures to ensure that they are sterile when placed on the market and remain sterile, under the storage and transport conditions laid down, until the protective packaging is damaged or opened.

8.4. Devices delivered in a sterile state must have been manufactured and sterilized by an appropriate, validated method.

8.5. Devices intended to be sterilized must be manufactured in appropriately controlled (e.g., environmental) conditions.

8.6. Packaging systems for non-sterile devices must keep the product without deterioration at the level of cleanliness stipulated and, if the devices are to be sterilized prior to use, minimize the risk of microbial contamination; the packaging system must be suitable taking account of the method of sterilization indicated by the manufacturer.

8.7. The packaging and/or label of the device must distinguish between identical or similar products sold in both sterile and non-sterile condition.

9. Construction and environmental properties

9.1. If the device is intended for use in combination with other devices or equipment, the whole combination, including the connection system must be safe and must not impair the specified performances of the devices. Any restrictions on use must be indicated on the label or in the instructions for use.

9.2. Devices must be designed and manufactured in such a way as to remove or minimize as far as is possible:

i. the risk of injury, in connection with their physical features, including the volume/ pressure ratio, dimensional, and where appropriate ergonomic features.

ii. risks connected with reasonably foreseeable environmental conditions, such as magnetic fields, external electrical influences, electrostatic discharge, pressure, temperature, or variations in pressure and acceleration.

iii. the risks of reciprocal interference with other devices normally used in the investigations or for the treatment given.

iv. risks arising where maintenance or calibration are not possible (as with implants), from aging of materials used or loss of accuracy of any measuring or control mechanism.

9.3. Devices must be designed and manufactured in such a way as to minimize the risks of fire or explosion during normal use and in single fault condition. Particular attention must be paid to devices whose intended use includes exposure to flammable substances or to substances which could cause combustion.

10. Devices with a measuring function

10.1. Devices with a measuring function must be designed and manufactured in such a way as to provide sufficient accuracy and stability within appropriate limits of accuracy and taking account of the intended purpose of the device. The limits of accuracy must be indicated by the manufacturer.

10.2. The measurement, monitoring, and display scale must be designed in line with ergonomic principles, taking account of the intended purpose of the device.

10.3. The measurements made by devices with a measuring function must be expressed in legal units conforming to the provisions of Council Directive 80/181/EEC (1).

11. Protection against radiation

11.1. General

11.1.1. Devices shall be designed and manufactured in such a way that exposure of patients, users, and other persons to radiation shall be reduced as far as possible compatible with the intended purpose, whilst not restricting the application of appropriate specified levels for therapeutic and diagnostic purposes.

11.2. Intended radiation

11.2.1. Where devices are designed to emit hazardous levels of radiation necessary for a specific medical purpose, the benefit of which is considered to outweigh the risks inherent in the emission, it must be possible for the user to control the emissions. Such devices shall be designed and manufactured to ensure reproducibility and tolerance of relevant variable parameters.

11.2.2. Where devices are intended to emit potentially hazardous, visible, and/or invisible radiation, they must be fitted, where practicable, with visual displays and/or audible warnings of such emissions.

11.3. Unintended radiation

11.3.1. Devices shall be designed and manufactured in such a way that exposure of patients, users, and other persons to the emission of unintended, stray or scattered radiation is reduced as far as possible.

11.4. Instructions

11.4.1. The operating instructions for devices emitting radiation must give detailed information as to the nature of the emitted radiation, means of protecting the patient and the user, and on ways of avoiding misuse and of eliminating the risks inherent in installation.

11.5. Ionizing radiation

11.5.1. Devices intended to emit ionizing radiation must be designed and manufactured in such a way as to ensure that, where practicable, the quantity, geometry, and quality of radiation emitted can be varied and controlled taking into account the intended use.

11.5.2. Devices emitting ionizing radiation intended for diagnostic radiology shall be designed and manufactured in such a way as to achieve appropriate image and/ or output quality for the intended medical purpose whilst minimizing radiation exposure of the patient and user.

11.5.3. Devices emitting ionizing radiation, intended for therapeutic radiology shall be designed and manufactured in such a way as to enable reliable monitoring and control of the delivered dose, the beam type and energy, and where appropriate the quality of radiation.

12. Requirements for medical devices connected to or equipped with an energy source

12.1. Devices incorporating electronic programmable systems must be designed to ensure the repeatability, reliability, and performance of these systems according to the intended use. In the event of a single fault condition (in the system), appropriate means should be adopted to eliminate or reduce as far as possible consequent risks.

12.1a. For devices which incorporate software or which are medical software in themselves, the software must be validated according to the state of the art, taking into account the principles of development lifecycle, risk management, validation, and verification.

12.2. Devices where the safety of the patients depends on an internal power supply must be equipped with a means of determining the state of the power supply.

12.3. Devices where the safety of the patients depends on an external power supply must include an alarm system to signal any power failure.

12.4. Devices intended to monitor one or more clinical parameters of a patient must be equipped with appropriate alarm systems to alert the user of situations which could lead to death or severe deterioration of the patient's state of health.

12.5. Devices must be designed and manufactured in such a way as to minimize the risks of creating electromagnetic fields which could impair the operation of other devices or equipment in the usual environment.

12.6. Protection against electrical risks

Devices must be designed and manufactured in such a way as to avoid, as far as possible, the risk of accidental electric shocks during normal use and in single fault condition, provided the devices are installed correctly.

12.7. Protection against mechanical and thermal risks

12.7.1. Devices must be designed and manufactured in such a way as to protect the patient and user against mechanical risks connected with, for example, resistance, stability, and moving parts.

12.7.2. Devices must be designed and manufactured in such a way as to reduce to the lowest possible level the risks arising from vibration generated by the devices, taking account of technical progress and of the means available for limiting vibrations, particularly at source, unless the vibrations are part of the specified performance.

12.7.3. Devices must be designed and manufactured in such a way as to reduce to the lowest possible level the risks arising from the noise emitted, taking account of technical progress and of the means available to reduce noise, particularly at source, unless the noise emitted is part of the specified performance.

12.7.4. Terminals and connectors to the electricity, gas or hydraulic and pneumatic energy supplies which the user has to handle must be designed and constructed in such a way as to minimize all possible risks.

12.7.5. Accessible parts of the devices (excluding the parts or areas intended to supply heat or reach given temperatures) and their surroundings must not attain potentially dangerous temperatures under normal use.

12.8. Protection against the risks posed to the patient by energy supplies or substances

12.8.1. Devices for supplying the patient with energy or substances must be designed and constructed in such a way that the flow rate can be set and maintained accurately enough to guarantee the safety of the patient and of the user

12.8.2. Devices must be fitted with the means of preventing and/or indicating any inadequacies in the flow rate which could pose a danger. Devices must incorporate suitable means to prevent, as far as possible, the accidental release of dangerous levels of energy from an energy and/or substance source.

12.9. The function of the controls and indicators must be clearly specified on the devices.

Where a device bears instruction required for its operation or indicates operating or adjustment parameters by means of a visual system, such information must be understandable to the user and, as appropriate, the patient.

13. Information supplied by the manufacturer

13.1. Each device must be accompanied by the information needed to use it safely and properly, taking account of the training and knowledge of the potential users, and to identify the manufacturer.

This information comprises the details on the label and the data in the instructions for use.

As far as practicable and appropriate, the information needed to use the device safely must be set out on the device itself and/or on the packaging for each unit or, where appropriate, on the sales packaging. If individual packaging of each unit is not practicable, the information must be set out in the leaflet supplied with one or more devices.

Instructions for use must be included in the packaging for every device. By way of exception, no such instructions for use are needed for devices in Class I or IIa if they can be used safely without any such instructions.

13.2. Where appropriate, this information should take the form of symbols. Any symbol or identification color used must conform to the harmonized standards. In areas for which no standards exist, the symbols and colors must be described in the documentation supplied with the device.

13.3. The label must bear the following particulars:

i. the name or trade name and address of the manufacturer. For devices imported into the Community, in view of their distribution in the Community, the label, or the outer packaging, or instructions for use, shall contain in addition the name and address of the authorized representative where the manufacturer does not have a registered place of business in the Community;

ii. the details strictly necessary to identify the device and the contents of the packaging especially for the users;

iii. where appropriate, the word 'STERILE';

iv.	where appropriate, the batch code, preceded by the word 'LOT,' or the serial number;
v.	where appropriate, an indication of the date by which the device should be used, in safety, expressed as the year and month;
vi.	where appropriate, an indication that the device is for single use. A manufacturer's indication of single-use must be consistent across the Community;
vii.	if the device is custom-made, the words 'custom-made device';
viii.	if the device is intended for clinical investigations, the words 'exclusively for clinical investigations;
ix.	any special storage and/or handling conditions;
x.	any special operating instructions;
xi.	any warnings and/or precautions to take;
xii.	year of manufacture for active devices other than those covered by (e). This indication may be included in the batch or serial number;
xiii.	where applicable, method of sterilization;
xiv.	in the case of a device within the meaning of Article 1(4a), an indication that the device contains a human blood derivative.
13.4.	If the intended purpose of the device is not obvious to the user, the manufacturer must clearly state it on the label and in the instructions for use.
13.5.	Wherever reasonable and practicable, the devices and detachable components must be identified, where appropriate in terms of batches, to allow all appropriate action to detect any potential risk posed by the devices and detachable components.
13.6.	Where appropriate, the instructions for use must contain the following particulars:
i.	the details referred to in Section 13.3, with the exception of (d) and (e);
ii.	the performances referred to in Section 3 and any undesirable side-effects;
iii.	if the device must be installed with or connected to other medical devices or equipment in order to operate as required for its intended purpose, sufficient details of its characteristics to identify the correct devices or equipment to use in order to obtain a safe combination;
iv.	all the information needed to verify whether the device is properly installed and can operate correctly and safely, plus details of the nature and frequency of the maintenance and calibration needed to ensure that the devices operate properly and safely at all times;
v.	where appropriate, information to avoid certain risks in connection with implantation of the device;
vi.	information regarding the risks of reciprocal interference posed by the presence of the device during specific investigations or treatment;
vii.	the necessary instructions in the event of damage to the sterile packaging and, where appropriate, details of appropriate methods of resterilization;
viii.	if the device is reusable, information on the appropriate processes to allow reuse, including cleaning, disinfection, packaging, and, where appropriate, the method of sterilization of the device to be resterilized, and any restriction on the number of reuses.

Where devices are supplied with the intention that they be sterilized before use, the instructions for cleaning and sterilization must be such that, if correctly followed, the device will still comply with the requirements in Section I.

If the device bears an indication that the device is for single-use, information on known characteristics and technical factors known to the manufacturer that could pose a risk if the device were to be re-used. If in accordance with Section 13.1 no instructions for use are needed, the information must be made available to the user upon request;

ix. details of any further treatment or handling needed before the device can be used (for example, sterilization, final assembly, etc.);

x. in the case of devices emitting radiation for medical purposes, details of the nature, type, intensity, and distribution of this radiation.

The instructions for use must also include details allowing the medical staff to brief the patient on any contra-indications and any precautions to be taken. These details should cover in particular:

xi. precautions to be taken in the event of changes in the performance of the device;

xii. precautions to be taken as regards exposure, in reasonably foreseeable environmental conditions, to magnetic fields, external electrical influences, electrostatic discharge, pressure or variations in pressure, acceleration, thermal ignition sources, etc.;

xiii. adequate information regarding the medicinal product or products which the device in question is designed to administer, including any limitations in the choice of substances to be delivered;

xiv. precautions to be taken against any special, unusual risks related to the disposal of the device;

xv. medicinal substances, or human blood derivatives incorporated into the device as an integral part in accordance with Section 7.4;

xvi. degree of accuracy claimed for devices with a measuring function;

xvii. date of issue or the latest revision of the instructions for use.

Compliance to the Essential Requirements is usually shown by completion of a check-list like documentation. Manufacturers are expected to indicate which requirements are applicable and which are not applicable. In addition, they are also required to show which standards they are in compliance with as well as what supporting evidence, they have to show this.

A simple template for manufacturers to use when completing their Essential Requirements is included in the appendix at the end of this volume. These ready-to-edit templates can be purchased from Scientists Sanctuary: https://scientistssanctuary.com/.

4.6. GENERAL SAFETY AND PERFORMANCE REQUIREMENTS (GSPR) UNDER THE MDR

While compliance with the Essential Requirements is imperative for establishing conformity with the MDD and active implantable medical device directive (AIMDD, 90/385/EEC), the same applies for the general safety and performance requirements (GSPR) within the MDR. The main purpose of this chapter is to compare the Essential requirements within the MDD to that of the GSPR of the MDR.

The first point to note is that the GSPRs are detailed in Annex I of the MDR and are structured into three chapters, namely:

- **Chapter I:** General requirements
- **Chapter II:** Requirements regarding design and manufacture
- **Chapter III:** Requirements regarding the information supplied with the device

Let's look at the requirements of Annex I (General Safety and Performance Requirements) in greater detail:

Annex I: General Safety and Performance Requirements
Chapter I: General Requirements
1. Devices shall achieve the performance intended by their manufacturer and shall be designed and manufactured in such a way that, during normal conditions of use, they are suitable for their intended purpose. They shall be safe and effective and shall not compromise the clinical condition or the safety of patients, or the safety and health of users or, where applicable, other persons, provided that any risks which may be associated with their use constitute acceptable risks when weighed against the benefits to the patient and are compatible with a high level of protection of health and safety, taking into account the generally acknowledged state of the art.
2. The requirement in this Annex to reduce risks as far as possible means the reduction of risks as far as possible without adversely affecting the benefit-risk ratio.
3. Manufacturers shall establish, implement, document, and maintain a risk management system.
Risk management shall be understood as a continuous iterative process throughout the entire lifecycle of a device, requiring regular systematic updating. In carrying out risk management manufacturers shall:
i. establish and document a risk management plan for each device;
ii. identify and analyze the known and foreseeable hazards associated with each device;
iii. estimate and evaluate the risks associated with, and occurring during, the intended use and during reasonably foreseeable misuse;
iv. eliminate or control the risks referred to in point (c) in accordance with the requirements of Section 4;

v. evaluate the impact of information from the production phase and, in particular, from the post-market surveillance system, on hazards and the frequency of occurrence thereof, on estimates of their associated risks, as well as on the overall risk, benefit-risk ratio and risk acceptability; and

vi. based on the evaluation of the impact of the information referred to in point (e), if necessary, amend control measures in line with the requirements of Section 4.

4. Risk control measures adopted by manufacturers for the design and manufacture of the devices shall conform to safety principles, taking account of the generally acknowledged state of the art. To reduce risks, Manufacturers shall manage risks so that the residual risk associated with each hazard as well as the overall residual risk is judged acceptable. In selecting the most appropriate solutions, manufacturers shall, in the following order of priority:

i. eliminate or reduce risks as far as possible through safe design and manufacture;

ii. where appropriate, take adequate protection measures, including alarms if necessary, in relation to risks that cannot be eliminated; and

iii. provide information for safety (warnings/precautions/contra-indications) and, where appropriate, training to users.

Manufacturers shall inform users of any residual risks.

5. In eliminating or reducing risks related to use error, the manufacturer shall:

i. reduce as far as possible the risks related to the ergonomic features of the device and the environment in which the device is intended to be used (design for patient safety); and

ii. give consideration to the technical knowledge, experience, education, training, and use environment, where applicable, and the medical and physical conditions of intended users (design for lay, professional, disabled or other users).

6. The characteristics and performance of a device shall not be adversely affected to such a degree that the health or safety of the patient or the user and, where applicable, of other persons are compromised during the lifetime of the device, as indicated by the manufacturer, when the device is subjected to the stresses which can occur during normal conditions of use and has been properly maintained in accordance with the manufacturer's instructions.

7. Devices shall be designed, manufactured, and packaged in such a way that their characteristics and performance during their intended use are not adversely affected during transport and storage, for example, through fluctuations of temperature and humidity, taking account of the instructions and information provided by the manufacturer.

8. All known and foreseeable risks, and any undesirable side-effects, shall be minimized and be acceptable when weighed against the evaluated benefits to the patient and/or user arising from the achieved performance of the device during normal conditions of use.

9. For the devices referred to in Annex XVI, the general safety requirements set out in Sections 1 and 8 shall be understood to mean that the device, when used under the conditions and for the purposes intended, does not present a risk at all or presents a risk that is no more than the maximum acceptable risk related to the product's use which is consistent with a high level of protection for the safety and health of persons.

Chapter II: Requirements Regarding Design and Manufacture

10. Chemical, physical, and biological properties

10.1. Devices shall be designed and manufactured in such a way as to ensure that the characteristics and performance requirements referred to in Chapter I are fulfilled. Particular attention shall be paid to:

i. the choice of materials and substances used, particularly as regards toxicity and, where relevant, flammability;

ii. the compatibility between the materials and substances used and biological tissues, cells, and body fluids, taking account of the intended purpose of the device and, where relevant, absorption, distribution, metabolism, and excretion;

iii. the compatibility between the different parts of a device which consists of more than one implantable part;

iv. the impact of processes on material properties;

v. where appropriate, the results of biophysical or modeling research the validity of which has been demonstrated beforehand;

vi. the mechanical properties of the materials used, reflecting, where appropriate, matters such as strength, ductility, fracture resistance, wear-resistance and fatigue resistance;

vii. surface properties; and

viii. the confirmation that the device meets any defined chemical and/or physical specifications.

10.2. Devices shall be designed, manufactured, and packaged in such a way as to minimize the risk posed by contaminants and residues to patients, taking account of the intended purpose of the device, and to the persons involved in the transport, storage, and use of the devices. Particular attention shall be paid to tissues exposed to those contaminants and residues and to the duration and frequency of exposure.

10.3. Devices shall be designed and manufactured in such a way that they can be used safely with the materials and substances, including gases, with which they enter into contact during their intended use; if the devices are intended to administer medicinal products, they shall be designed and manufactured in such a way as to be compatible with the medicinal products concerned in accordance with the provisions and restrictions governing those medicinal products and that the performance of both the medicinal products and of the devices is maintained in accordance with their respective indications and intended use.

10.4. Substances

10.4.1. Design and manufacture of devices

Devices shall be designed and manufactured in such a way as to reduce as far as possible the risks posed by substances or particles, including wear debris, degradation products, and processing residues, that may be released from the device.

Devices, or those parts thereof or those materials used therein that:

i. are invasive and come into direct contact with the human body.

ii. (re)administer medicines, body-liquids or other substances, including gases, to/from the body, or

iii. transport or store such medicines, body fluids or substances, including gases, to be (re)administered to the body, shall only contain the following substances in a concentration that is above 0.1% weight by weight (w/w) where justified pursuant to Section 10.4.2:

a. substances which are carcinogenic, mutagenic or toxic to reproduction ('CMR'), of category 1A or 1B, in accordance with Part 3 of Annex VI to Regulation (EC) No 1272/2008 of the European Parliament and of the Council (1), or

b. substances having endocrine-disrupting properties for which there is scientific evidence of probable serious effects to human health and which are identified either in accordance with the procedure set out in Article 59 of Regulation (EC) No 1907/2006 of the European Parliament and of the Council (2) or, once a delegated act has been adopted by the Commission pursuant to the first subparagraph of Article 5(3) of Regulation (EU) No 528/2012 of the European Parliament and the Council (3), in accordance with the criteria that are relevant to human health amongst the criteria established therein.

10.4.2. Justification regarding the presence of CMR and/or endocrine-disrupting substances

The justification for the presence of such substances shall be based upon:

i. an analysis and estimation of potential patient or user exposure to the substance;

ii. an analysis of possible alternative substances, materials or designs, including, where available, information about independent research, peer-reviewed studies, scientific opinions from relevant scientific committees and an analysis of the availability of such alternatives;

iii. argumentation as to why possible substance and/or material substitutes, if available, or design changes, if feasible, are inappropriate in relation to maintaining the functionality, performance, and the benefit-risk ratios of the product; including taking into account if the intended use of such devices includes treatment of children or treatment of pregnant or breastfeeding women or treatment of other patient groups considered particularly vulnerable to such substances and/or materials; and

iv. where applicable and available, the latest relevant scientific committee guidelines in accordance with Sections 10.4.3. and 10.4.4.

10.4.3. Guidelines on phthalates

For the purposes of Section 10.4., the Commission shall, as soon as possible and by 26 May 2018, provide the relevant scientific committee with a mandate to prepare guidelines that shall be ready before 26 May 2020. The mandate for the committee shall encompass at least a benefit-risk assessment of the presence of phthalates which belong to either of the groups of substances referred to in points (a) and (b) of Section 10.4.1. The benefit-risk assessment shall take into account the intended purpose and context of the use of the device, as well as any available alternative substances and alternative materials, designs, or medical treatments. When deemed appropriate on the basis of the latest scientific evidence, but at least every five years, the guidelines shall be updated.

10.4.4. Guidelines on other CMR and endocrine-disrupting substances

Subsequently, the Commission shall mandate the relevant scientific committee to prepare guidelines as referred to in Section 10.4.3. also for other substances referred to in points (a) and (b) of Section 10.4.1., where appropriate.

10.4.5. Labeling

Where devices, parts thereof or materials used therein as referred to in Section 10.4.1. contain substances referred to in points (a) or (b) of Section 10.4.1. in a concentration above 0.1% weight by weight (w/w), the presence of those substances shall be labeled on the device itself and/or on the packaging for each unit or, where appropriate, on the sales packaging, with the list of such substances. If the intended use of such devices includes treatment of children or treatment of pregnant or breastfeeding women or treatment of other patient groups considered particularly vulnerable to such substances and/or materials, information on residual risks for those patient groups and, if applicable, on appropriate precautionary measures shall be given in the instructions for use.

10.5. Devices shall be designed and manufactured in such a way as to reduce as far as possible the risks posed by the unintentional ingress of substances into the device, taking into account the device and the nature of the environment in which it is intended to be used.

10.6. Devices shall be designed and manufactured in such a way as to reduce as far as possible the risks linked to the size and the properties of particles which are or can be released into the patient's or user's body, unless they come into contact with intact skin only. Special attention shall be given to nanomaterials.

11. Infection and microbial contamination

11.1. Devices and their manufacturing processes shall be designed in such a way as to eliminate or to reduce as far as possible the risk of infection to patients, users, and, where applicable, other persons. The design shall:

i. reduce as far as possible and appropriate the risks from unintended cuts and pricks, such as needle stick injuries.

ii. allow easy and safe handling.

iii. reduce as far as possible any microbial leakage from the device and/or microbial exposure during use, and

iv. prevent microbial contamination of the device or its content such as specimens or fluids.

11.2. Where necessary devices shall be designed to facilitate their safe cleaning, disinfection, and/or re-sterilization.

11.3. Devices labeled as having a specific microbial state shall be designed, manufactured, and packaged to ensure that they remain in that state when placed on the market and remain so under the transport and storage conditions specified by the manufacturer.

11.4. Devices delivered in a sterile state shall be designed, manufactured, and packaged in accordance with appropriate procedures, to ensure that they are sterile when placed on the market and that, unless the packaging which is intended to maintain their sterile condition is damaged, they remain sterile, under the transport and storage conditions specified by the manufacturer, until that packaging is opened at the point of use. It shall be ensured that the integrity of that packaging is clearly evident to the final user.

11.5. Devices labeled as sterile shall be processed, manufactured, packaged, and sterilized by means of appropriate, validated methods.

11.6. Devices intended to be sterilized shall be manufactured and packaged in appropriate and controlled conditions and facilities.

11.7. Packaging systems for non-sterile devices shall maintain the integrity and cleanliness of the product and, where the devices are to be sterilized prior to use, minimize the risk of microbial contamination; the packaging system shall be suitable taking account of the method of sterilization indicated by the manufacturer.

11.8. The labeling of the device shall distinguish between identical or similar devices placed on the market in both a sterile and a non-sterile condition additional to the symbol used to indicate that devices are sterile.

12. Devices incorporating a substance considered to be a medicinal product and devices that are composed of substances or of combinations of substances that are absorbed by or locally dispersed in the human body.

12.1. In the case of devices referred to in the first subparagraph of Article 1(8), the quality, safety, and usefulness of the substance which, if used separately, would be considered to be a medicinal product within the meaning of point (2) of Article 1 of Directive 2001/83/EC, shall be verified by analogy with the methods specified in Annex I to Directive 2001/83/EC, as required by the applicable conformity assessment procedure under this Regulation.

12.2. Devices that are composed of substances or of combinations of substances that are intended to be introduced into the human body, and that are absorbed by or locally dispersed in the human body shall comply, where applicable and in a manner limited to the aspects not covered by this Regulation, with the relevant requirements laid down in Annex I to Directive 2001/83/EC for the evaluation of absorption, distribution, metabolism, excretion, local tolerance, toxicity, interaction with other devices, medicinal products or other substances and potential for adverse reactions, as required by the applicable conformity assessment procedure under this Regulation.

13. Devices incorporating materials of biological origin

13.1. For devices manufactured utilizing derivatives of tissues or cells of human origin which are non-viable or are rendered non-viable covered by this Regulation in accordance with point (g) of Article 1(6), the following shall apply:

i. donation, procurement, and testing of the tissues and cells shall be done in accordance with Directive 2004/23/EC;

ii. processing, preservation, and any other handling of those tissues and cells or their derivatives shall be carried out so as to provide safety for patients, users, and, where applicable, other persons. In particular, safety with regard to viruses and other transmissible agents shall be addressed by appropriate methods of sourcing and by implementation of validated methods of elimination or inactivation in the course of the manufacturing process;

iii. the traceability system for those devices shall be complementary and compatible with the traceability and data protection requirements laid down in Directive 2004/23/EC and in Directive 2002/98/EC.

13.2. For devices manufactured utilizing tissues or cells of animal origin, or their derivatives, which are non-viable or rendered non-viable the following shall apply:

i. where feasible, taking into account the animal species, tissues, and cells of animal origin, or their derivatives, shall originate from animals that have been subjected to veterinary controls that are adapted to the intended use of the tissues. Information on the geographical origin of the animals shall be retained by manufacturers;

ii. sourcing, processing, preservation, testing, and handling of tissues, cells, and substances of animal origin, or their derivatives, shall be carried out so as to provide safety for patients, users, and, where applicable, other persons. In particular safety with regard to viruses and other transmissible agents shall be addressed by implementation of validated methods of elimination or viral inactivation in the course of the manufacturing process, except when the use of such methods would lead to unacceptable degradation compromising the clinical benefit of the device;

iii. in the case of devices manufactured utilizing tissues or cells of animal origin, or their derivatives, as referred to in Regulation (EU) No 722/2012 the particular requirements laid down in that Regulation shall apply.

13.3. For devices manufactured utilizing non-viable biological substances other than those referred to in Sections 13.1 and 13.2, the processing, preservation, testing, and handling of those substances shall be carried out so as to provide safety for patients, users, and, where applicable, other persons, including in the waste disposal chain. In particular, safety with regard to viruses and other transmissible agents shall be addressed by appropriate methods of sourcing and by implementation of validated methods of elimination or inactivation in the course of the manufacturing process.

14. Construction of devices and interaction with their environment

14.1. If the device is intended for use in combination with other devices or equipment, the whole combination, including the connection system shall be safe and shall not impair the specified performance of the devices. Any restrictions on use applying to such combinations shall be indicated on the label and/or in the instructions for use. Connections which the user has to handle, such as fluid, gas transfer, electrical or mechanical coupling, shall be designed and constructed in such a way as to minimize all possible risks, such as misconnection.

14.2. Devices shall be designed and manufactured in such a way as to remove or reduce as far as possible:

i. the risk of injury, in connection with their physical features, including the volume/pressure ratio, dimensional, and where appropriate ergonomic features;

ii. risks connected with reasonably foreseeable external influences or environmental conditions, such as magnetic fields, external electrical and electromagnetic effects, electrostatic discharge, radiation associated with diagnostic or therapeutic procedures, pressure, humidity, temperature, variations in pressure and acceleration or radio signal interferences;

iii. the risks associated with the use of the device when it comes into contact with materials, liquids, and substances, including gases, to which it is exposed during normal conditions of use;

iv. the risks associated with the possible negative interaction between software and the IT environment within which it operates and interacts;

v. the risks of accidental ingress of substances into the device;

vi. the risks of reciprocal interference with other devices normally used in the investigations or for the treatment given; and

vii. risks arising where maintenance or calibration are not possible (as with implants), from aging of materials used or loss of accuracy of any measuring or control mechanism.

14.3. Devices shall be designed and manufactured in such a way as to minimize the risks of fire or explosion during normal use and in single fault condition. Particular attention shall be paid to devices thc intended use of which includes exposure to or use in association with flammable or explosive substances or substances which could cause combustion.

14.4. Devices shall be designed and manufactured in such a way that adjustment, calibration, and maintenance can be done safely and effectively.

14.5. Devices that are intended to be operated together with other devices or products shall be designed and manufactured in such a way that the interoperability and compatibility are reliable and safe.

14.6. Any measurement, monitoring, or display scale shall be designed and manufactured in line with ergonomic principles, taking account of the intended purpose, users, and the environmental conditions in which the devices are intended to be used.

14.7. Devices shall be designed and manufactured in such a way as to facilitate their safe disposal and the safe disposal of related waste substances by the user, patient or other person. To that end, manufacturers shall identify and test procedures and measures as a result of which their devices can be safely disposed of after use. Such procedures shall be described in the instructions for use.

15. Devices with a diagnostic or measuring function

15.1. Diagnostic devices and devices with a measuring function, shall be designed and manufactured in such a way as to provide sufficient accuracy, precision, and stability for their intended purpose, based on appropriate scientific and technical methods. The limits of accuracy shall be indicated by the manufacturer.

15.2. The measurements made by devices with a measuring function shall be expressed in legal units conforming to the provisions of Council Directive 80/181/EEC (1).

16.	Protection against radiation
16.1.	General
i.	Devices shall be designed, manufactured, and packaged in such a way that exposure of patients, users, and other persons to radiation is reduced as far as possible, and in a manner that is compatible with the intended purpose, whilst not restricting the application of appropriate specified levels for therapeutic and diagnostic purposes.
ii.	The operating instructions for devices emitting hazardous or potentially hazardous radiation shall contain detailed information as to the nature of the emitted radiation, the means of protecting the patient and the user, and on ways of avoiding misuse and of reducing the risks inherent to installation as far as possible and appropriate. Information regarding the acceptance and performance testing, the acceptance criteria, and the maintenance procedure shall also be specified.
16.2.	Intended radiation
i.	Where devices are designed to emit hazardous, or potentially hazardous, levels of ionizing and/or non-ionizing radiation necessary for a specific medical purpose, the benefit of which is considered to outweigh the risks inherent to the emission, it shall be possible for the user to control the emissions. Such devices shall be designed and manufactured to ensure reproducibility of relevant variable parameters within an acceptable tolerance.
ii.	Where devices are intended to emit hazardous, or potentially hazardous, ionizing, and/or non-ionizing radiation, they shall be fitted, where possible, with visual displays and/or audible warnings of such emissions.
16.3.	Devices shall be designed and manufactured in such a way that exposure of patients, users, and other persons to the emission of unintended, stray or scattered radiation is reduced as far as possible. Where possible and appropriate, methods shall be selected which reduce the exposure to radiation of patients, users, and other persons who may be affected.
16.4.	Ionizing radiation
i.	Devices intended to emit ionizing radiation shall be designed and manufactured taking into account the requirements of the Directive 2013/59/Euratom laying down basic safety standards for protection against the dangers arising from exposure to ionizing radiation.
ii.	Devices intended to emit ionizing radiation shall be designed and manufactured in such a way as to ensure that, where possible, taking into account the intended use, the quantity, geometry, and quality of the radiation emitted can be varied and controlled, and, if possible, monitored during treatment.
iii.	Devices emitting ionizing radiation intended for diagnostic radiology shall be designed and manufactured in such a way as to achieve an image and/or output quality that are appropriate to the intended medical purpose whilst minimizing radiation exposure of the patient and user.
iv.	Devices that emit ionizing radiation and are intended for therapeutic radiology shall be designed and manufactured in such a way as to enable reliable monitoring and control of the delivered dose, the beam type, energy, and, where appropriate, the quality of radiation.

17.	Electronic programmable systems—devices that incorporate electronic programmable systems and software that are devices in themselves
17.1.	Devices that incorporate electronic programmable systems, including software, or software that are devices in themselves, shall be designed to ensure repeatability, reliability, and performance in line with their intended use. In the event of a single fault condition, appropriate means shall be adopted to eliminate or reduce as far as possible consequent risks or impairment of performance.
17.2.	For devices that incorporate software or for software that are devices in themselves, the software shall be developed and manufactured in accordance with the state of the art, taking into account the principles of development life cycle, risk management, including information security, verification, and validation.
17.3.	Software referred to in this Section that is intended to be used in combination with mobile computing platforms shall be designed and manufactured taking into account the specific features of the mobile platform (e.g., size, and contrast ratio of the screen) and the external factors related to their use (varying environment as regards level of light or noise).
17.4.	Manufacturers shall set out minimum requirements concerning hardware, IT networks characteristics and IT security measures, including protection against unauthorized access, necessary to run the software as intended.
18.	Active devices and devices connected to them
18.1.	For non-implantable active devices, in the event of a single fault condition, appropriate means shall be adopted to eliminate or reduce as far as possible consequent risks.
18.2.	Devices where the safety of the patient depends on an internal power supply shall be equipped with a means of determining the state of the power supply and an appropriate warning or indication for when the capacity of the power supply becomes critical. If necessary, such warning or indication shall be given prior to the power supply becoming critical.
18.3.	Devices where the safety of the patient depends on an external power supply shall include an alarm system to signal any power failure.
18.4.	Devices intended to monitor one or more clinical parameters of a patient shall be equipped with appropriate alarm systems to alert the user of situations which could lead to death or severe deterioration of the patient's state of health.
18.5.	Devices shall be designed and manufactured in such a way as to reduce as far as possible the risks of creating electromagnetic interference which could impair the operation of the device in question or other devices or equipment in the intended environment.
18.6.	Devices shall be designed and manufactured in such a way as to provide a level of intrinsic immunity to electromagnetic interference that is adequate to enable them to operate as intended.
18.7.	Devices shall be designed and manufactured in such a way as to avoid, as far as possible, the risk of accidental electric shocks to the patient, user or any other person, both during normal use of the device and in the event of a single fault condition in the device, provided the device is installed and maintained as indicated by the manufacturer.

18.8.	Devices shall be designed and manufactured in such a way as to protect, as far as possible, against unauthorized access that could hamper the device from functioning as intended.
19.	Particular requirements for active implantable devices
19.1.	Active implantable devices shall be designed and manufactured in such a way as to remove or minimize as far as possible:
i.	risks connected with the use of energy sources with particular reference, where electricity is used, to insulation, leakage currents and overheating of the devices.
ii.	risks connected with medical treatment, in particular those resulting from the use of defibrillators or high-frequency surgical equipment, and
iii.	risks which may arise where maintenance and calibration are impossible, including:
a.	excessive increase of leakage currents.
b.	aging of the materials used.
c.	excess heat generated by the device.
d.	decreased accuracy of any measuring or control mechanism.
19.2.	Active implantable devices shall be designed and manufactured in such a way as to ensure
i.	if applicable, the compatibility of the devices with the substances they are intended to administer, and
ii.	the reliability of the source of energy.
19.3.	Active implantable devices and, if appropriate, their component parts shall be identifiable to allow any necessary measure to be taken following the discovery of a potential risk in connection with the devices or their component parts.
19.4.	Active implantable devices shall bear a code by which they and their manufacturer can be unequivocally identified (particularly with regard to the type of device and its year of manufacture); it shall be possible to read this code, if necessary, without the need for a surgical operation.
20.	Protection against mechanical and thermal risks
20.1.	Devices shall be designed and manufactured in such a way as to protect patients and users against mechanical risks connected with, for example, resistance to movement, instability, and moving parts.
20.2.	Devices shall be designed and manufactured in such a way as to reduce to the lowest possible level the risks arising from vibration generated by the devices, taking account of technical progress and of the means available for limiting vibrations, particularly at source, unless the vibrations are part of the specified performance.
20.3.	Devices shall be designed and manufactured in such a way as to reduce to the lowest possible level the risks arising from the noise emitted, taking account of technical progress and of the means available to reduce noise, particularly at source, unless the noise emitted is part of the specified performance.
20.4.	Terminals and connectors to the electricity, gas or hydraulic and pneumatic energy supplies which the user or other person has to handle, shall be designed and constructed in such a way as to minimize all possible risks.

20.5. Errors likely to be made when fitting or refitting certain parts which could be a source of risk shall be made impossible by the design and construction of such parts or, failing this, by information given on the parts themselves and/or their housings.

The same information shall be given on moving parts and/or their housings where the direction of movement needs to be known in order to avoid a risk.

20.6. Accessible parts of devices (excluding the parts or areas intended to supply heat or reach given temperatures) and their surroundings shall not attain potentially dangerous temperatures under normal conditions of use.

21. Protection against the risks posed to the patient or user by devices supplying energy or substances

21.1. Devices for supplying the patient with energy or substances shall be designed and constructed in such a way that the amount to be delivered can be set and maintained accurately enough to ensure the safety of the patient and of the user.

21.2. Devices shall be fitted with the means of preventing and/or indicating any inadequacies in the amount of energy delivered or substances delivered which could pose a danger. Devices shall incorporate suitable means to prevent, as far as possible, the accidental release of dangerous levels of energy or substances from an energy and/or substance source.

21.3. The function of the controls and indicators shall be clearly specified on the devices. Where a device bears instructions required for its operation or indicates operating or adjustment parameters by means of a visual system, such information shall be understandable to the user and, as appropriate, the patient.

22. Protection against the risks posed by medical devices intended by the manufacturer for use by laypersons

22.1. Devices for use by laypersons shall be designed and manufactured in such a way that they perform appropriately for their intended purpose taking into account the skills and the means available to laypersons and the influence resulting from variation that can be reasonably anticipated in the layperson's technique and environment. The information and instructions provided by the manufacturer shall be easy for the layperson to understand and apply.

22.2. Devices for use by laypersons shall be designed and manufactured in such a way as to:

i. ensure that the device can be used safely and accurately by the intended user at all stages of the procedure, if necessary after appropriate training and/or information.

ii. reduce, as far as possible and appropriate, the risk from unintended cuts and pricks such as needle stick injuries, and

iii. reduce as far as possible the risk of error by the intended user in the handling of the device and, if applicable, in the interpretation of the results.

22.3. Devices for use by laypersons shall, where appropriate, include a procedure by which the layperson:

i. can verify that, at the time of use, the device will perform as intended by the manufacturer, and

ii. if applicable, is warned if the device has failed to provide a valid result.

While there are 13 Essential Requirements in the MDD and 16 in the AIMDD, there are 23 general safety and performance requirements (GSPRs) in the new MDR. The overall requirements have been expanded, but the scope is the same as that of the previous directives. There are a few notable exceptions. For example, clinical evaluation has moved into the articles section of the regulation, and a few new topics have been added, such as devices without a medical purpose and requirements for devices used by laypersons. In addition, some parts of the GSPR now have increased emphasis and more detailed requirements so as to allow for more thorough alignment with harmonized standards and industry guidance. Some numbering and organizational changes have taken place, the requirements for risk and labeling have been expanded, there are new areas of emphasis as per standards and industry guidance, and many "state of the art" requirements from harmonized standards have been incorporated directly into the Regulation's GSPRs.

Let's take a closer look at some of the detail of the GSPR:

- **GSPR 1: Performance and Safety:** GSPR 1 generally corresponds to MDD ER 1 with the majority of the text being the same. This requirement states that the devices shall be 'designed and manufactured in such a way' that the safety of patients and users shall not be compromised. As with ER 1 in the directives, this is under the normal conditions of use. The concept of 'performance' is also brought in from the MDD ER 3. Design and construction of the device should conform to safety principles, taking into account the 'generally acknowledged state of the art' as required in to MDD ER 2. Risks relating to ergonomic features and consideration of the use environment, which was previously present in MDD ER 1 have now moved to GSPR 5.
- **GSPR 2: Reduction of Risks:** The GSPR 2–5 all relate to risk management. GSPR 2 is a new statement addressing the requirement to reduce risks as far as possible without adversely affecting the risk-benefit ratio.
- **GSPR 3: Risk Management System:** GSPR 3 is a new requirement which more explicitly defines the requirements of risk management as aligned with EN ISO 14971.
- **GSPR 4: Risk Control Measures and Residual Risks:** GSPR 4 generally corresponds to ER 2 in the existing MDD and covers the requirements that both, the overall residual risks *and* the residual risk associated with each hazard is evaluated and judged to be acceptable, with respect to the benefits. While this is already

an expectation of the harmonized standard, this will now be explicitly included in the Regulation. Furthermore, there is also a requirement that the manufacturer 'shall inform users of any residual risks.'

- **GSPR 5: Risks Related to Use:** GSPR 5 corresponds to the latter part of ER 1 in the MDD and addresses the reduction of risks relating to use error. Usability or human factors are addressed such that the risk of use error shall be reduced as far as possible.
- **GSPR 6: Device Lifetime:** GSPR 6 defines requirements for the characteristics and device performance over the lifetime of the device as indicated. The device characteristics and performance shall not be adversely impacted to such a degree that health or safety of a patient or user would be compromised, when under the stresses of 'normal conditions of use.' Device performance must not be adversely impacted when the device 'has been properly maintained in accordance with the manufacturer's instructions.' Reasonably foreseeable maintenance or storage errors may be considered as part of use risk evaluation, but in some cases, impacts to performance cannot be prevented when maintenance and storage are not per the manufacturer's instructions.
- **GSPR 7: Packaging, Transport, Storage:** This requirement outlines the basics for device packaging, transportation, and storage. This requirement can be linked to ER 5 in the MDD. The regulation provides examples of potential temperature and humidity fluctuations which are commonly considered by manufacturers during transportation simulations and storage where necessary.
- **GSPR 8: Risk-Benefit Ratio:** Here an updated definition of the 'risk-benefit ratio' is provided. It now includes 'all known and foreseeable risks,' and states that the risks 'shall be minimized.'
- **GSPR 9: Devices Without a Medical Purpose:** Devices without a medical purpose are excluded from the scope of the MDD but are included in the MDR. GSPR 9 clarifies how to apply the 'risk-benefit' and 'performance' requirements for those devices without a medical purpose or medical benefit.
- **GSPR 10: Chemical, Physical, and Biological Properties:** The requirements of GSPR 10 are broadly related to biological safety with more focus on the areas of specific substances of concern.

- **GSPR 11: Infection and Microbial Contamination:** The requirements of GSPR 11 are generally similar to that of MDD ER 8 and relate to infection and microbial contamination. This includes risk of infection, design for reuse, devices with a specific microbial state, devices delivered sterile, validation for sterile devices, environmental controls, packaging for non-sterile devices and labeling for sterile state.
- **GSPR 12: Devices Incorporating a Medicinal Product, Substances Absorbed or Locally Dispersed:** GSPR 12 deals with the requirements for medicinal products or substances which are absorbed or locally dispersed in the body. Medicinal substances were previously addressed in MDD ER 7.4. It should be noted that the scope of GSPR 12 has significantly changed under MDR with a broader scope with regards to the definition of what is included along with the medicinal considerations which now includes 'substances or combinations of substances that are absorbed by or locally dispersed in the human body.'
- **GSPR 13: Devices Incorporating Materials of Biological Origin:** Materials of biological origin within the MDR has expanded a great deal from what was originally included in the MDD. These requirements generally correspond to MDD ER 7.4 and 8.2 and address tissues, cells or derivatives of human origin.
- **GSPR 14: Construction of Devices and Interaction with Their Environment:** GSPR 14 corresponds to ER 9 in the MDD as well as ER 10.2 for ergonomic principles and deals with the construction of devices and interaction with their environment. Specifically, it focuses on use in combination, risks of interaction with the environment, risks of fire or explosion, design for adjustment, calibration, and maintenance; design for compatibility, measurement, monitoring or display scales and design and manufacture for safe disposal.
- **GSPR 15: Devices with a Diagnostic or Measuring Function:** GSPR 15 of the MDR (in comparison to the MDD) deals with 'diagnostic devices' which has been added to the scope of the requirement. Manufacturers should consider the 'precision' of the device and measurements from measuring devices are still required to be expressed in legal units as per Directive 80/181/EEC.

- **GSPR 16: Protection Against Radiation:** These requirements in GSPR 16 generally correspond to parts of ER 11 in the MDD and covers protection against radiation. This includes general requirements, requirements for intended and unintended radiation as well as ionizing radiation.
- **GSPR 17: Electronic Programmable Systems and Software:** GSPR 17 defines the requirements for electronic programmable systems and software considered to be a medical device. The specific wording of the Regulation now includes 'software that are devices in themselves' more explicitly in the scope of these requirements. Information security are also included.
- **GSPR 18: Active Devices and Devices Connected to Them:** GSPR 18 defines particular requirements for active devices and connected devices with a large portion of the requirement corresponding to subparts of ER 12 in the MDD.
- **GSPR 19: Particular Requirements for Active Implantable Devices:** GSPR 19 defines the requirements for active implantable devices. The requirements can be generally linked to ERs in the current AIMDD; however as the MDD and AIMDD requirements have now been merged into the single MDR, many of the requirements for active implantable devices have moved and are listed here. This includes particular risks to be reduced for active implantable devices, device compatibility and reliability of energy, identification of devices and components, and identification codes.
- **GSPR 20: Protection Against Mechanical and Thermal Risks:** The requirements in GSPR 20 can be directly correlated to ER 12.7 in the MDD. This includes errors that could be made in fitting or refitting parts. This is usually covered as manufacturers align to the current standards and usability requirements.
- **GSPR 21: Protection Against the Risks Posed to the Patient or User by Devices Supplying Energy or Substances:** GSPR 21 corresponds to ERs 12.8.1, 12.8.2 and 12.9 in the MDD and deals with protection against the risks posed to the patient or user by devices supplying energy or substances. There is some minor rewording of the requirement however the intent is unchanged.
- **GSPR 22: Protection Against the Risks Posed by Medical Devices Intended by the Manufacturer for Use by Lay Persons:**

Specific requirements for devices used by laypersons are new in the MDR. GSPR 22 outlines the requirements and considerations for the device's instructions, variation in user technique, risks of error and injury, and verification of device function by the user.

- **GSPR 23: Label and Instructions for Use (IFU):** GSPR 23 deals with product labels and IFU in which the text and requirements for labels and IFUs are greatly expanded as compared with both Directives. This includes the general requirements for information supplied by the manufacturer, label requirements as well as sterile package label requirements and requirements pertaining to the instructions for use (IFU).

Chapter 5

Technical File/STED

CONTENTS

Without proper formatting and organization, even the most detailed medical device applications can fail to pass regulatory review, in turn, costing manufacturers time and large amounts of money. The summary technical documentation (STED) format was created by the GHTF in an effort to globally standardize medical device regulatory submissions across markets. STED is recognized by the US, European, Canadian, Australian, and Japanese regulators to name a few. The GHTF proposed a harmonized format for the STED for Medical Devices, in which all relevant information should be presented. Depending on the classification of the device, the information contained in the summary may include abstracts, summaries, or existing controlled documents, as well as an essential principles (EP) checklist.

5.1. WHAT IS THE PURPOSE OF STED?

Manufacturers of all classes of medical devices are expected to demonstrate conformity of the device to the EP of Safety and Performance. This is done through the preparation of technical documentation that shows how the device was developed, designed, and manufactured. For the purpose of a conformity assessment, the manufacturer must assemble the STED from existing technical documentation. The STED reflects the status of the device at a particular moment in time and is prepared to show the device meets regulatory requirements. The content of the STED is the same for both pre- and post-market use but it should be noted that the circumstances for the use of the STED are different. The STED should be prepared in a language acceptable to the reviewing organization, and the overall depth and detail of the information contained in the STED will depend on the classification of the device itself. Ideally, the STED should contain summary information on selected topics while also containing detailed information on certain other topics (elaborated later on) as well as an EP checklist. The EP checklist is created as part of the manufacturer's technical documentation and is controlled by the manufacturer's QMS. It provides a tabular overview of the EP and identifies those that are applicable to the device as well as the chosen method of demonstrating that the device conforms to each relevant Essential Principle and the reference of the controlled document that is relevant to a specific Essential Principle.

5.2. THE USE OF STED IN THE PREMARKET PHASE

In the premarket phase, the STED will be prepared and submitted to the RA/CAB for Class C and D IVD medical devices. For Class A and B IVD medical devices, the STED will be prepared and submitted only at the request of a RA/CAB. In this instance, the manufacturer should be able to assemble and submit it in the timeframe indicated by the RA/CAB. It should also be remembered that content of any submitted STED should be traceable by the manufacturer for future reference.

5.3. THE USE OF THE STED IN THE POST-MARKET PHASE

In the post-market phase, the RA/CAB may request submission of a STED to investigate the continued conformity of a particular device. If requested, the manufacturer should be able to prepare and submit the STED in the timeframe indicated by the RA/CAB. The content of any submitted STED should be traceable by the manufacturer for future reference. Ideally, the manufacturer should be able to gain and review data about a device which is already being marketed by use of a post-marketing surveillance system.

5.4. CONTENTS OF STED

5.4.1. Device Description Including Variants and Accessories

5.4.1.1. Device Description

The STED should include the following information in the device description section:

- The intended use of the IVD medical device. This section may include:
- what is detected
- its function such as screening, monitoring, diagnosis or aid to diagnosis
- the specific disorder, condition or risk factor of interest that it is intended to detect, define or differentiate
- whether it is automated or not
- whether it is qualitative or quantitative

- the type of specimen(s) required
- Testing population
- The intended user (layperson or professional);
- A general description of the principle of the assay method or instrument principles of operation
- The Class of the device and the applicable classification rule according to *Principles of In Vitro Diagnostic (IVD) Medical Devices Classification*
- A description of the components and where appropriate, a description of the reactive ingredients of other relevant components such as antibodies, antigens, nucleic acid primers, etc., and where applicable:
- A description of the specimen collection and transport materials provided with the IVD medical device or descriptions of specifications recommended for use
- For instruments of automated assays: a description of the appropriate assay characteristics or dedicated assays
- For automated assays: a description of the appropriate instrumentation characteristics or dedicated instrumentation;
- A description of any software to be used with the IVD medical device;
- A description or complete list of the various configurations/ variants of the IVD medical device that will be made available
- a description of the accessories, other IVD medical devices and other products that are not IVD medical devices, which are intended to be used in combination with the IVD medical device.

5.4.2. Reference to the Manufacturer's Previous Device Generation(s) and/or Similar Devices or Device History

5.4.2.1. For an IVD Medical Device Not Yet Available on Any Market

The STED may provide a summary of:

- The manufacturer's previous generation(s) of the IVD medical device, if such exists; and

- The manufacturer's similar IVD medical devices available on the market.

5.4.2.2. For an IVD Medical Device Already Available on the Market in Any Jurisdiction

This information may include a summary of the number of adverse event reports related to the safety and performance of this IVD medical device in relation to the number of IVD medical devices placed on the market. External certificates and documents which give written evidence of conformity with the EP may be annexed to the STED.

5.4.3. Essential Principles (EP) Checklist

The STED should include an EP checklist that identifies:

- the Essential Principles;
- for each Essential Principle whether it applies to the IVD medical device and if not, why not;
- the method used to demonstrate conformity with each Essential Principle that applies; and
- the reference to the actual technical documentation that offers evidence of conformity with each method used.

The method used to demonstrate conformity may include one or more of the following:

- conformity with recognized or other standards;
- conformity with a commonly accepted industry test method;
- conformity with appropriate in-house test methods that have been validated and verified; and
- Comparison to an IVD medical device already available on the market.

The EP checklist should include a cross-reference to the location of such evidence both within the full technical documentation held by the manufacturer and within the STED itself. A template of an EP Checklist as laid out in the *Summary Technical Documentation for Demonstrating Conformity to the EP of Safety and Performance of Medical Devices (STED)*-GHTF 2007 is given below.

5.4.4. How to Fill Out an EP Checklist

1. **Identity of the Medical Device:** The manufacturer should identify the device, and when applicable the various configurations/variants covered by the checklist
2. **Applicability to the Device:** Is the listed Essential Principle applicable to the device? Here the answer is either 'Yes' or 'No.' If the answer is 'No', this should be briefly explained.
3. **Method Used to Demonstrate Conformity:** In this column, the manufacturer should state the type(s) of method(s) that it has chosen to demonstrate conformity, e.g., the recognized standard(s), industry or in-house test method(s), comparison study(ies) or other method used.
4. **Method Reference:** After having stated the method in the previous column, here the manufacturer should name the title and should reference the recognized standard(s), industry or in-house test method(s), comparison study or studies or other method used to demonstrate conformity. For standards, this should include the date of the standard and where appropriate, the clause(s) that demonstrates conformity with the relevant EP.
5. **Reference to Supporting Controlled Documents:** This column should contain a reference to the actual technical documentation that demonstrates conformity to the Essential Principle, i.e., the certificates, test reports, study reports or other documents that resulted from the method used to demonstrate conformity and, if included in the STED, its location.

Essential Principal Checklist	
Device:	

Essential Principal	**Applicable to the Device?**	**Method of Conformity**	**Identity of Specific Documents**
General Requirements			

5.1. Medical devices should be designed and manufactured in such a way that, when used under the conditions and for the purposes intended and, where applicable, by virtue of the technical knowledge, experience, education or training of intended users, they will not compromise the clinical condition or the safety of patients, or the safety and health of users or, where applicable, other persons, provided that any risks which may be associated with their use constitute acceptable risks when weighed against the benefits to the patient and are compatible with a high level of protection of health and safety.			
5.2. The solutions adopted by the manufacturer for the design and manufacture of the devices should conform to safety principles, taking account of the generally acknowledged state of the art. When risk reduction is required, the manufacturer should control the risk(s) so that the residual risk(s) associated with each hazard is judged acceptable. The manufacturer should apply the following principles in the priority order listed: • identify known or foreseeable hazards and estimate the associated risks arising from the intended use and foreseeable misuse. • eliminate risks as far as reasonably practicable through inherently safe design and manufacture. • reduce as far as is reasonably practicable the remaining risks by taking adequate protection measures, including alarms. • Inform users of any residual risks.			

5.3. Devices should achieve the performance intended by the manufacturer and be designed, manufactured, and packaged in such a way that they are suitable for one or more of the functions within the scope of the definition of a medical device applicable in each jurisdiction.			
5.4. The characteristics and performances referred to in Clauses 5.1, 5.2 and 5.3 should not be adversely affected to such a degree that the health or safety of the patient or the user and, where applicable, of other persons are compromised during the lifetime of the device, as indicated by the manufacturer, when the device is subjected to the stresses which can occur during normal conditions of use and has been properly maintained in accordance with the manufacturer's instructions.			
5.5. The devices should be designed, manufactured, and packed in such a way that their characteristics and performances during their intended use will not be adversely affected under transport and storage conditions (for example, fluctuations of temperature and humidity) taking account of the instructions and information provided by the manufacturer.			
5.6. The benefits must be determined to outweigh any undesirable side effects for the performances intended.			
Design and Manufacturing Requirements			
5.7. Chemical, physical, and biological properties			

5.7.1. The devices should be designed and manufactured in such a way as to ensure the characteristics and performance referred to in Clauses 5.1 to 5.6 of the 'General Requirements.' Particular attention should be paid to: • the choice of materials used, particularly as regards toxicity and, where appropriate, flammability. • the compatibility between the materials used and biological tissues, cells, body fluids, and specimens, taking account of the intended purpose of the device. • The choice of materials used should reflect, where appropriate, matters such as hardness, wear, and fatigue strength.			
5.7.2. The devices should be designed, manufactured, and packed in such a way as to minimize the risk posed by contaminants and residues to the persons involved in the transport, storage, and use of the devices and to patients, taking account of the intended purpose of the product. Particular attention should be paid to tissues exposed and to the duration and frequency of exposure.			
5.7.3. The devices should be designed and manufactured in such a way that they can be used safely with the materials, substances, and gases with which they enter into contact during their normal use or during routine procedures; if the devices are intended to administer medicinal products they should be designed and manufactured in such a way as to be compatible with the medicinal products concerned according to the provisions and restrictions governing these products and that their performance is maintained in accordance with the intended use.			

5.7.4. Where a device incorporates, as an integral part, a substance which, if used separately, may be considered to be a medicinal product/drug as defined in the relevant legislation that applies within that jurisdiction and which is liable to act upon the body with action ancillary to that of the device, the safety, quality, and usefulness of the substance should be verified, taking account of the intended purpose of the device.			
5.7.5. The devices should be designed and manufactured in such a way as to reduce as far as reasonably practicable and appropriate the risks posed by substances that may leach or leak from the device.			
5.7.6. Devices should be designed and manufactured in such a way as to reduce as far as reasonably practicable and appropriate risks posed by the unintentional ingress or egress of substances into or from the device taking into account the device and the nature of the environment in which it is intended to be used.			
5.8. Infection and microbial contamination			

5.8.1. The devices and manufacturing processes should be designed in such a way as to eliminate or to reduce as far as reasonably practicable and appropriate the risk of infection to patients, users, and, where applicable, other persons. The design should: • allow easy handling, and, where necessary: • reduce as far as reasonably practicable and appropriate any microbial leakage from the device and/or microbial exposure during use. • Prevent microbial contamination of the device, or specimen where applicable, by the patient, user or other person.			
5.8.2. Where a device incorporates substances of biological origin, the risk of infection must be reduced as far as reasonably practicable and appropriate by selecting appropriate sources, donors, and substances and by using, as appropriate, validated inactivation, conservation, and test and control procedures.			

5.8.3. In some jurisdictions products incorporating tissues, cells, and substances of non-human origin may be considered medical devices. In this case, such tissues, cells, and substances should originate from animals that have been subjected to veterinary controls and surveillance adapted to the intended use of the tissues. National regulations may require that the manufacturer and/or the Regulatory Authority retain information on the geographical origin of the animals. Processing, preservation, testing, and handling of tissues, cells, and substances of animal origin should be carried out so as to provide optimal safety. In particular, safety with regard to viruses and other transmissible agents should be addressed by implementation of validated methods of elimination or inactivation in the course of the manufacturing process.			
In some jurisdictions, products incorporating human tissues, cells, and substances may be considered medical devices. In this case, the selection of sources, donors, and/or substances of human origin, the processing, preservation, testing, and handling of tissues, cells, and substances of such origin should be carried out so as to provide optimal safety. In particular, safety with regard to viruses and other transmissible agents should be addressed by implementation of validated methods of elimination or inactivation in the course of the manufacturing process.			
Devices labeled as having a special microbiological state should be designed, manufactured, and packed to ensure they remain so when placed on the market and remain so under the transport and storage conditions specified by the manufacturer.			

Devices delivered in a sterile state should be designed, manufactured, and packed in a non-reusable pack, and/or according to appropriate procedures, to ensure that they are sterile when placed on the market and remain sterile, under the transport and storage conditions indicated by the manufacturer, until the protective packaging is damaged or opened.			
5.8.4. Devices labeled either as sterile or as having a special microbiological state should have been processed, manufactured, and, if applicable, sterilized by appropriate, validated methods.			
5.8.5. Devices intended to be sterilized should be manufactured in appropriately controlled (e.g., environmental) conditions.			
5.8.6. Packaging systems for non-sterile devices should keep the product without deterioration at the level of cleanliness stipulated and, if the devices are to be sterilized prior to use, minimize the risk of microbial contamination; the packaging system should be suitable taking account of the method of sterilization indicated by the manufacturer.			
5.8.7. The packaging and/or label of the device should distinguish between identical or similar products placed on the market in both sterile and non-sterile condition.			
5.9. Manufacturing and Environmental Properties			

5.9.1. If the device is intended for use in combination with other devices or equipment, the whole combination, including the connection system should be safe and should not impair the specified performance of the devices. Any restrictions on use applying to such combinations should be indicated on the label and/or in the instructions for use.			
5.9.2. Devices should be designed and manufactured in such a way as to remove or reduce as far as reasonably practicable and appropriate: • the risk of injury, in connection with their physical features, including the volume/pressure ratio, dimensional, and where appropriate ergonomic features; • risks connected with reasonably forcsceable external influences or environmental conditions, such as magnetic fields, external electrical and electromagnetic effects, electrostatic discharge, pressure, humidity, temperature or variations in pressure and acceleration; • the risks connected to their use in conjunction with materials, substances, and gases with which they may come into contact during normal conditions of use; • the risks of accidental penetration of substances into the device; • the risk of incorrect identification of specimens; • the risks of reciprocal interference with other devices normally used in the investigations or for the treatment given; • Risks arising where maintenance or calibration are not possible (as with implants), from aging of materials used or loss of accuracy of any measuring or control mechanism.			

5.9.3. Devices should be designed and manufactured in such a way as to minimize the risks of fire or explosion during normal use and in single fault condition. Particular attention should be paid to devices whose intended use includes exposure to or use in association with flammable substances or substances which could cause combustion.			
5.9.4. Devices must be designed and manufactured in such a way as to facilitate the safe disposal of any waste substances.			
5.10. Devices with a Diagnostic or Measuring Function			
5.10.1. Devices with a measuring function, where inaccuracy could have a significant adverse effect on the patient, should be designed and manufactured in such a way as to provide sufficient accuracy, precision, and stability for their intended purpose of the device. The limits of accuracy should be indicated by the manufacturer.			
5.10.2. Diagnostic devices should be designed and manufactured in such a way as to provide sufficient accuracy, precision, and stability for their intended use, based on appropriate scientific and technical methods. In particular, the design should address sensitivity, specificity, trueness, repeatability, and reproducibility, control of known relevant interference and limits of detection, as appropriate.			
5.10.3. Where the performance of devices depends on the use of calibrators and/or control materials, the traceability of values assigned to such calibrators and/or control materials should be assured through a quality management system.			

5.10.4. Any measurement, monitoring or display scale should be designed in line with ergonomic principles, taking account of the intended purpose of the device.			
5.10.5. Wherever possible values expressed numerically should be in commonly accepted, standardized units, and understood by the users of the device. *Note:* While SG1 generally supports convergence on the global use of internationally standardized measurement units, considerations of safety, user familiarity, and established clinical practice may justify the use of other recognized measurement units.			
5.11. Protection against radiation			
5.11.1. General			
5.11.1.1. Devices should be designed and manufactured and packaged in such a way that exposure of patients, users, and other persons to any emitted radiation should be reduced as far as practicable and appropriate, compatible with the intended purpose, whilst not restricting the application of appropriate specified levels for therapeutic and diagnostic purposes.			
5.11.2. Intended radiation			
5.11.2.1. Where devices are designed to emit hazardous, or potentially hazardous, levels of visible and/or invisible radiation necessary for a specific medical purpose, the benefit of which is considered to outweigh the risks inherent in the emission, it should be possible for the user to control the emissions. Such devices should be designed and manufactured to ensure reproducibility of relevant variable parameters within an acceptable tolerance.			

5.11.2.2. Where devices are intended to emit potentially hazardous, visible, and/or invisible radiation, they should be fitted, where practicable, with visual displays and/or audible warnings of such emissions.			
5.11.3. Unintended radiation			
5.11.3.1. Devices should be designed and manufactured in such a way that exposure of patients, users, and other persons to the emission of unintended, stray or scattered radiation is reduced as far as practicable and appropriate.			
5.11.4. Instructions for use			
5.11.4.1. The operating instructions for devices emitting radiation should give detailed information as to the nature of the emitted radiation, means of protecting the patient and the user, and on ways of avoiding misuse and of eliminating the risks inherent in installation.			
5.11.5. Ionizing radiation			
5.11.5.1. Devices intended to emit ionizing radiation should be designed and manufactured in such a way as to ensure that, where practicable, the quantity, geometry, and energy distribution (or quality) of radiation emitted can be varied and controlled taking into account the intended use.			
5.11.5.2. Devices emitting ionizing radiation intended for diagnostic radiology should be designed and manufactured in such a way as to achieve appropriate image and/or output quality for the intended medical purpose whilst minimizing radiation exposure of the patient and user.			

5.11.5.3. Devices emitting ionizing radiation, intended for therapeutic radiology should be designed and manufactured in such a way as to enable reliable monitoring and control of the delivered dose, the beam type, and energy and where appropriate the energy distribution of the radiation beam.			
5.12. Requirements for medical devices connected to or equipped with an energy source			
5.12.1. Devices incorporating electronic programmable systems, including software, should be designed to ensure the repeatability, reliability, and performance of these systems according to the intended use. In the event of a single fault condition in the system, appropriate means should be adopted to eliminate or reduce as far as practicable and appropriate consequent risks.			
5.12.2. Devices where the safety of the patients depends on an internal power supply should be equipped with a means of determining the state of the power supply.			
5.12.3. Devices where the safety of the patients depends on an external power supply should include an alarm system to signal any power failure.			
5.12.4. Devices intended to monitor one or more clinical parameters of a patient should be equipped with appropriate alarm systems to alert the user of situations which could lead to death or severe deterioration of the patient's state of health			

5.12.5. Devices should be designed and manufactured in such a way as to reduce as far as practicable and appropriate the risks of creating electromagnetic interference which could impair the operation of this or other devices or equipment in the usual environment.			
5.12.6. Devices should be designed and manufactured in such a way as to provide an adequate level of intrinsic immunity to electromagnetic disturbance to enable them to operate as intended.			
5.12.7. Protection against electrical risks Devices should be designed and manufactured in such a way as to avoid, as far as possible, the risk of accidental electric shocks during normal use and in single fault condition, provided the devices are installed and maintained as indicated by the manufacturer.			
5.13. Protection against mechanical risks			
5.13.1. Devices should be designed and manufactured in such a way as to protect the patient and user against mechanical risks connected with, for example, resistance to movement, instability, and moving parts.			
5.13.2. Devices should be designed and manufactured in such a way as to reduce to the lowest practicable level the risks arising from vibration generated by the devices, taking account of technical progress and of the means available for limiting vibrations, particularly at source, unless the vibrations are part of the specified performance.			

5.13.3. Devices should be designed and manufactured in such a way as to reduce to the lowest practicable level the risks arising from the noise emitted, taking account of technical progress and of the means available to reduce noise, particularly at source, unless the noise emitted is part of the specified performance.			
5.13.4. Terminals and connectors to the electricity, gas or hydraulic and pneumatic energy supplies which the user has to handle should be designed and constructed in such a way as to minimize all possible risks.			
5.13.5. Accessible parts of the devices (excluding the parts or areas intended to supply heat or reach given temperatures) and their surroundings should not attain potentially dangerous temperatures under normal use.			
5.14. Protection against the risks posed to the patient by supplied energy or substances			
5.14.1. Devices for supplying the patient with energy or substances should be designed and constructed in such a way that the delivered amount can be set and maintained accurately enough to guarantee the safety of the patient and of the user.			
5.14.2. Devices should be fitted with the means of preventing and/or indicating any inadequacies in the delivered amount which could pose a danger. Devices should incorporate suitable means to prevent, as far as possible, the accidental release of dangerous levels of energy from an energy and/or substance source.			

5.14.3. The function of the controls and indicators should be clearly specified on the devices. Where a device bears instructions required for its operation or indicates operating or adjustment parameters by means of a visual system, such information should be understandable to the user and, as appropriate, the patient.			
5.15. Protection against the risks posed to the patient for devices for self-testing or self-administration			
5.15.1. Such devices should be designed and manufactured in such a way that they perform appropriately for their intended purpose taking into account the skills and the means available to users and the influence resulting from variation that can reasonably be anticipated in user's technique and environment. The information and instructions provided by the manufacturer should be easy for the user to understand and apply.			
5.15.2. Such devices should be designed and manufactured in such a way as to reduce as far as practicable the risk of use error in the handling of the device and, if applicable, the specimen, and also in the interpretation of results.			
5.15.3. Such devices should, where reasonably possible, include a procedure by which the user can verify that, at the time of use, that the product will perform as intended by the manufacturer.			
5.16. Information supplied by the manufacturer			

5.16.1. Users should be provided with the information needed to identify the manufacturer, to use the device safely and to ensure the intended performance, taking account of their training and knowledge. This information should be easily understood. *Note:* Further information is provided in SG1/N009 Labeling for Medical Devices and in SG1/N043 Labeling for Medical Devices (revised).			
5.17. Performance evaluation including, where appropriate, clinical evaluation			
5.17.1. All data generated in support of performance evaluation should be obtained in accordance with the relevant requirements applicable in each jurisdiction.			
5.17.2. Clinical investigations on human subjects should be carried out in accordance with the spirit of the Helsinki Declaration. This includes every step in the clinical investigation from the first consideration of the need and justification of the study to publication of the results. In addition, some countries may have specific regulatory requirements for pre-study protocol review or informed consent.			

5.4.5. Risk Analysis and Control Summary

The STED should contain a summary of the risks identified during the risk analysis process and a description of how these risks have been controlled to an acceptable level. Ideally, this risk analysis should be based on recognized standards and be part of the manufacturer's overall risk management plan and risk file. The summary should address possible hazards for the device and any indirect risks which may result from device-associated hazards. The results of the risk analysis should provide a conclusion with evidence that any remaining risks are acceptable when compared to the benefits.

5.4.6. Design and Manufacturing Information

5.4.6.1. Device Design

The device design section should contain substantial information to allow a reviewer to obtain a general understanding of the design applied to the device. It should include a description of the critical components or ingredients as well. For instruments and software, an overview of the entire system would be required, including a flowchart of the relationships among the major functional units in the software such as relationships to hardware and so on.

5.4.6.2. Manufacturing Processes

This is generally only applicable to Class D devices. The manufacturing processes section of the STED should contain information that allows a reviewer to obtain a general understanding of the manufacturing processes. The information may take the form of a process flow chart showing an overview of production, including the technologies used as well as assembly and packaging of the finished medical device. This section should include details of any in-process and final product testing as well.

5.4.6.3. Manufacturing Sites

The STED should identify the sites where various manufacturing activities are performed (ideally this would include the sites that are involved in critical manufacturing activities). If QMS certificates, or the equivalent, exist for these sites, they may be referenced in the STED and included in an annex.

5.4.7. Product Verification and Validation

The information provided in the product verification and validation section of the STED will vary in the level of detail as determined by the classification of the device. In addition, other characteristics will influence the level of detail provided. As the STED should summarize the results of verification and validation studies undertaken to demonstrate conformity of the device with the EP that apply to it. Where appropriate, such information might come from the literature.

For the purpose of the STED document, 'summary' and 'detailed information' are defined in details as follows:

1. **Summary Information:** A summary should provide enough information to allow the RA/CAB to assess the validity of that

information and should contain a brief description of:

- the study protocol;
- the study results;
- the study conclusion.

This summary may also include:

- Where a recognized standard exists: a declaration/certificate of conformity to a recognized standard can be provided with a summary of the data if no acceptance criteria are specified in the standard;
- In the absence of a recognized standard: a declaration/certificate of conformity to a published standard that has not been recognized might be provided if it is supported by a rationale for its use, and summary of the data, and a conclusion, if no acceptance criteria are specified in the standard;
- In the absence of a recognized standard and non-recognized published standards: a professional guideline, industry method, or in-house standard may be referred to in the summarized information. However, it should be supported by a rationale for its use, a description of the method used, a summary of the data in sufficient detail and a conclusion to allow assessment of its adequacy; and
- A review of relevant published literature regarding the device or substantially similar devices.

2. **Detailed Information:** This should include:

- the complete study protocol;
- the method of data analysis;
- the complete study report; and
- the study conclusion.

Where a recognized standard exists that contains the protocol and the method of data analysis, this information can be substituted by a declaration/certificate of conformity to the recognized standard. However, a summary of the data and conclusions should also be provided. For clinical performance, the detailed information will typically include individual data points for a Class D device. Where appropriate, actual test result summaries with their acceptance criteria should be provided and not just pass/fail statements.

5.4.8. Analytical Performance

The statements and descriptions in the following sections refer to all IVD medical devices. There are applicability differences between instrumentation and reagent-based assays. The essays themselves may be quantitative, semi-quantitative or qualitative in nature. There may be limited applicability of some of the following subsections for qualitative or semi-quantitative assays.

1. **Analytical Performance Characteristics: Specimen Type:** This section should describe the different specimen types that can be used and should include their stability and storage conditions. It is typically applicable to all systems and assay types. Stability includes storage (such as duration, temperature limits, and freeze/thaw cycles) and, where applicable, transportation conditions. This section should include summary information for each matrix and anticoagulant as well as a description of the measurement procedure for comparison or determination of measurement accuracy. This includes information such as specimen type tested, number of samples, sample range or target concentrations tested, calculations, and statistical methods, results, and conclusions. Higher risk class devices should contain more detailed information.
2. **Analytical Performance Characteristics: Accuracy of measurement:** This section should describe both trueness and precision studies. Measurement trueness is normally expressed in terms of bias, while measurement precision is naturally expressed in terms of standard deviation. Accuracy is affected by a combination of systematic and random effects that contribute as individual components of the total error of measurement.
3. **Analytical Performance Characteristics: Trueness of Measurement:** This section should provide information on the trueness of the measurement procedure and summarize the data in sufficient detail to allow assessment of the adequacy of the means selected to establish the trueness. Trueness measures apply to both quantitative and qualitative assays only when a reference standard or method is available. Higher risk class devices (Class C and D) should contain more detailed information.
4. **Analytical Performance Characteristics: Precision of Measurement:** This section should describe repeatability and reproducibility studies.

5. **Analytical Performance Characteristics: Repeatability:** This section should include repeatability estimates and information about the studies used to estimate variability within runs. Repeatability data is obtained for instrumentation in conjunction with an appropriate assay. Higher risk class devices (Class C and D) should contain more detailed information. These studies should include the use of samples that represent the full range of expected analyze concentrations that can be measured by the test as claimed by the manufacturer. If a recognized standard is used, a declaration/certificate of conformity to the recognized standard along with a summary of the data and conclusions should be provided.
6. **Analytical Performance Characteristics: Reproducibility:** This section should include reproducibility estimates and information about the studies used to estimate variability between days, runs, sites, lots, operators, and instruments. Such variability is also known as 'Intermediate Precision." Higher risk class devices (Class C and D) should contain more detailed information.
7. **Analytical Performance Characteristics: Analytical Sensitivity:** This section should include information about the study design and results. It should provide a description of the specimen type and preparation including matrix, analyte levels, and how levels were established. The number of replicates tested at each concentration should also be provided as well as a description of the calculation used to determine assay sensitivity. Higher risk class devices (Class C and D) should contain more detailed information.
8. **Analytical Performance Characteristics: Analytical Specificity:** This section should describe interference and cross-reactivity studies to determine the analytical specificity, defined as the ability of a measurement procedure to detect or measure only the analyte to be detected, in the presence of other substances/agents in the sample. This section should also aim to provide information on the evaluation of potentially interfering and cross-reacting substances/agents on the assay. Information should be provided on the substance/agent type and concentration tested, sample type, analyze test concentration, and results. Higher risk class devices (Class C and D) should contain more detailed information.

9. **Analytical Performance Characteristics: Metrological Traceability of Calibrator and Control Material Values:** Where applicable, this section should aim to summarize information about metrological traceability of values assigned to calibrators and trueness control materials. Methods and acceptance criteria for the metrological traceability to reference materials and/or reference measurement procedures should be included as well as a description of value assignment and validation. Higher risk class devices (Class C and D) should contain more detailed information.
10. **Analytical Performance Characteristics: Measuring Range of the Assay:** This section should include a summary of studies that define the measuring range of the assay. It should include the limit of detection and describe information on how these were established. It should further include a description of specimen type, number of samples, number of replicates, and preparation, including information on matrix, analyze levels and how levels were established. Higher risk class devices (Class C and D) should contain more detailed information.
11. **Analytical Performance Characteristics: Definition of Assay Cut-Off:** This section should provide a summary of analytical data with a description of the study design, including methods for determining the assay cut-off levels. This can include:
    i. The population(s) studied (demographics/selection/inclusion and exclusion criteria/number of individuals included);
    ii. Method or mode of characterization of specimens;
    iii. Statistical methods used to generate results; and
    iv. If applicable, define gray zones/equivocal zones.

Higher risk class devices (Class C and D) should contain more detailed information.

1. **Clinical Performance:** Ideally, the STED should contain data on the clinical performance of the device such that it demonstrates conformity of the device to the EP that apply to it.
2. **Stability:** This section should provide a summarized description on claimed shelf life, in use stability and shipping studies.
    i. **Claimed Shelf Life:** This section should provide information on stability testing studies to support the duration of shelf life claimed by the manufacturer. Testing should be performed on at

least three different lots manufactured under conditions that are essentially equivalent to routine production conditions. These lots do not need to be consecutive lots. Accelerated studies or extrapolated data from real-time data are acceptable for initial shelf life claim but need to be followed up with real-time stability studies. Higher risk class devices (Class C and D) should contain more detailed information.

Details included in this section should include:

a. The study report detailing the protocol, number of lots, acceptance criteria and testing intervals;

b. When accelerated studies have been performed in anticipation of the real-time studies, the method used for accelerated studies; and

c. Conclusions and claimed shelf life.

Note: Shelf life can be derived from the lot with the longest real-time stability data as long as accelerated or extrapolated data from all three lots are comparable.

ii. **In-Use Stability:** This section should provide information on in use stability studies for one lot reflecting actual routine use of the device (real or simulated). This may include open vial stability and/or, for automated instruments, on board stability. Where automated instrumentation is used and if calibration stability is claimed, then supporting data should be included.

Details included in this section should include:

a. The study report detailing the protocol, acceptance criteria and testing intervals; and

b. Conclusions and claimed in use stability.

Higher risk class devices (Class C and D) should contain more detailed information.

iii. **Shipping Stability:** This section should provide information on shipping stability studies for one lot to evaluate the tolerance of products to the anticipated shipping conditions.

Shipping studies can be done under real and/or simulated conditions and should include variable shipping conditions such as extreme heat and/or cold.

Details included in this section should include:

a. The study report which includes the protocol and related acceptance criteria;

b. The method used for simulated conditions; and
c. Conclusion and recommended shipping conditions.

Higher risk class devices (Class C and D) should contain more detailed information.

5.4.9. Software Verification and Validation

Evidence of the validation of the software as used in the finished device should be provided. This information should typically include a summary of results pertaining to verification, validation, and testing performed in-house as well as that which is applicable in an actual user environment prior to final release. It should also address all of the different hardware configurations and, where applicable, operating systems identified in the labeling.

Higher risk class devices (Class C and D) should contain more detailed information.

5.4.10. Labeling

A complete set of labeling associated with the device should be provided, and this information should be aligned with that which is provided in the GHTF Guideline on Labeling for Medical Devices (GHTF/SG1/N43:2005 Labeling for Medical Devices).

This information provided should include:

- All labels associated with the device, both immediate and the outer container; and
- Instructions for use.

Ideally, where the STED is submitted to a RA/CAB, then the labeling set of documents should be translated into a language as required by the relevant regulatory authority, in addition, the labeling documents should contain the final content as determined by the manufacturer. This, however, does not have to be in the final (printed) format.

5.4.11. Format of STED

Ideally, the STED should be organized such that it incorporates the same sections as described in the guidance document laid out by the GHTF. Manufacturers should make reference to GHTF/SG1/SG1/N063:2011: Summary technical documentation (STED) for Demonstrating Conformity to the EP of Safety and Performance of IVD Medical Devices (STED).

5.4.12. Declaration of Conformity (DoC)

The declaration of conformity (DoC) does not form part of the STED however; it may be annexed to the STED once the conformity assessment process has been completed. Manufacturers should make reference to GHTF/SG1/N46:2007: Principles of Conformity Assessment for IVD Medical Devices.

Chapter 6

Appointing an Authorized Representative

The European Authorized Representative serves as the liaison between the manufacturer and the national competent authorities or Ministries of Health in the EU. A European Authorized Representative may also be known as the EUAR, EC Rep, CE Rep, EU Rep and/or an EAR. All of the abovementioned abbreviations mean the same thing: European Authorized Representative. The European Authorized Representative serves as the point of contact between the manufacturer (which is a non-European medical device company) and the National Competent Authorities and notified bodies (NB).

The roles and responsibilities of the authorized representative include:

- Operating in accordance with MEDDEV 2.5/10 Guidelines for authorized representatives (AR).
- Assisting with medical device registrations within the European Union (EU).
- Holding a copy of the manufacturer's technical file and/or CE DoC in case it is requested by the relevant National Competent Authority (CA).
- Assisting with Incident and field safety corrective actions (FSCA) reporting as per MEDDEV 2.12/1.
- Provide, if requested, a regulatory review of product artwork and labeling.
- Provide regulatory news and updates.

The name and address of the Authorized Representative must be included on the labeling, outer packaging and instructions for use (IFU) and the appropriate EC Rep symbol must also be used as per EN ISO 15223-1:2016.

It is possible to also appoint a distributor as an Authorized Representative in Europe; however, this is usually not recommended because most distributors are not prepared to fulfill the mandatory responsibilities of the EC Representative. Manufacturers should instead opt to select an independent representative that focuses directly on regulatory affairs.

It is also important to note that under the new MDR 2017/745, AR will have greater responsibilities and will take on significantly more risk and liabilities. These responsibilities are often laid out in the agreement between an Authorized Representative and the manufacturer.

Chapter 7 Declaration of Conformity

The DoC is a document prepared by the manufacturer stating that the medical device complies fully with all regulatory requirements. As a minimum, this declaration should contain the following information:

- An attestation that each device that is subject to the declaration complies with the applicable EP for Safety and Performance and the applicable requirements of label and instructions for use (IFU) for Medical Devices.
- Information sufficient to identify the device/s to which the DoC applies.
- The global medical device nomenclature (GMDN) code for the device.
- The classification of the device/s after following the guidance found in principles of medical devices classification.
- The date on which the DoC is issued.
- The name and address of the device manufacturer.
- The name, position, and signature of the responsible person who has been authorized to complete the DoC upon the manufacturer's behalf.

The Regulatory Authority or Conformity Assessment Body may review and confirm the adequacy of the DoC and, if required, examine the supporting documents or other evidence.

Chapter

8

The Medical Device Directive 93/42/EEC

CONTENTS

Any medical device placed on the market must comply with the relevant legislation of the country which that device is entering. For Europe, this was the medical device directive (MDD)-in transition to the medical device regulation (MDR). This means, for a manufacturer to legally place a medical device on the European market, they must show that their device meets the essential requirements of the relevant regulation.

The legislation consists of three directives:

- Directive 90/385/EEC for active implantable medical devices (AIMD);
- Directive 93/42/EEC for medical devices (MDD); and
- Directive 98/79/EC for *in vitro* diagnostic medical devices (IVDD).

As such, all devices being sold within the EU were required to conform to the directive by June 1, 1998. It's important to understand that the main purpose of the MDD was to harmonize those standards that benefit manufacturers, users, and patients while also defining the specific requirements for clinical testing, design, manufacture, testing/inspection, marketing, installation, and service of medical devices being sold within the EU.

The MDD consists of 23 articles, 12 annexes and 18 classification rules. Within the MDD, three main areas of focus for medical devices were:

- the essential requirements;
- classification rules; and
- conformity routes for assessments.

The Medical Devices Directive was enforced to provide a harmonized regulatory environment for all medical devices sold within the EU. All products which fall within the scope of the Directive must meet certain essential safety and administrative requirements before being CE marked to show compliance. Such products may then be freely sold within the EU without being subject to any additional national regulations.

8.1. SCOPE OF THE MDD

The Medical Devices Directive is one of three directives which together cover all medical equipment. The remaining two directives are: The Active Implantable Medical Devices Directive (AIMDD) and the *In Vitro* Diagnostic Devices Directive (IVDD).

Let's look at how the Directive defines medical devices. According to the MDD: a medical device is defined as: "any instrument, apparatus, appliance, material or other article, whether used alone or in combination, including the software necessary for its proper application intended by the manufacturer to be used for human beings for the purpose of:

- diagnosis, prevention, monitoring, treatment or alleviation of disease.
- diagnosis, monitoring, treatment, or alleviation of or compensation for an injury or handicap.
- investigation, replacement or modification of the anatomy or of a physiological process.
- control of conception and which does not achieve its principal intended action in or on the human body by pharmacological, immunological or metabolic means, but which may be assisted in its function by such means."

The scope of the MDD is quite wide, and in some instances, there may be some form of overlap with other Directives. The MDD does however include provisions for all relevant hazards such as mechanical and electrical safety as well as provisions for electromagnetic compatibility. Devices which fall within the scope of the MDD, the AIMDD and the IVDD are generally excluded from the scope of other Directives. It is also important for a manufacturer to realize that some devices/equipment may fall into or out of the scope of the MDD solely on the basis of the purpose to which the equipment is intended to be used and the way it is marketed. In simpler terms, if a manufacturer makes any claims that a piece of equipment has some form of therapeutic effect or which alleviates the symptoms of any disability, injury or illness, then it falls within the scope of the MDD, and a product registration is required as well as appropriate documentation. Common examples of products which are considered to be medical devices include:

- Medicine measuring cups;
- Syringes;
- Dental instruments;
- Stethoscopes;
- Thermometers;
- Prescription spectacles and contact lenses;
- Bandages and splints;

- Dental treatment chairs;
- Wheelchairs;
- Condoms;
- First aid kits.

8.2. PRODUCTS COVERED BY THE MDD

Article 1 of the MDD defines a 'medical device' to mean any instrument, apparatus, appliance, software, material or other article, whether used alone or in combination, including the software intended by its manufacturer to be used specifically for diagnostic and/or therapeutic purposes and necessary for its proper application, intended by the manufacturer to be used for human beings for the purpose of:

- diagnosis, prevention, monitoring, treatment or alleviation of disease;
- diagnosis, monitoring, treatment, alleviation of or compensation for an injury or handicap;
- investigation, replacement or modification of the anatomy or of a physiological process;
- control of conception; and which does not achieve its principal intended action in or on the human body by pharmacological, immunological or metabolic means, but which may be assisted in its function by such means;

In addition, the scope includes an 'accessory,' which means an article which whilst not being a device is intended specifically by its manufacturer to be used together with a device to enable it to be used.

Article 1 of the MDD 93/42/EEC: Definitions, Scope	
1.	This Directive shall apply to medical devices and their accessories. For the purposes of this Directive, accessories shall be treated as medical devices in their own right. Both medical devices and accessories shall hereinafter be termed devices.
2.	For the purposes of this Directive, the following definitions shall apply:
i.	'medical device' means any instrument, apparatus, appliance, software, material or other article, whether used alone or in combination, including the software intended by its manufacturer to be used specifically for diagnostic and/or therapeutic purposes and necessary for its proper application, intended by the manufacturer to be used for human beings for the purpose of:

a. diagnosis, prevention, monitoring, treatment or alleviation of disease.

b. diagnosis, monitoring, treatment, alleviation of or compensation for an injury or handicap.

c. investigation, replacement or modification of the anatomy or of a physiological process.

d. control of conception, and which does not achieve its principal intended action in or on the human body by pharmacological, immunological or metabolic means, but which may be assisted in its function by such means;

ii. 'accessory' means an article which whilst not being a device is intended specifically by its manufacturer to be used together with a device to enable it to be used in accordance with the use of the device intended by the manufacturer of the device;

iii. '*in vitro* diagnostic medical device' means any medical device which is a reagent, reagent product, calibrator, control material, kit, instrument, apparatus, equipment or system, whether used alone or in combination, intended by the manufacturer to be used *in vitro* for the examination of specimens, including blood and tissue donations, derived from the human body, solely or principally for the purpose of providing information:

a. concerning a physiological or pathological state, or

b. concerning a congenital abnormality, or

c. to determine the safety and compatibility with potential recipients, or

d. to monitor therapeutic measures.

Specimen receptacles are considered to be *in vitro* diagnostic medical devices. 'Specimen receptacles' are those devices, whether vacuum-type or not, specifically intended by their manufacturers for the primary containment and preservation of specimens derived from the human body for the purpose of *in vitro* diagnostic examination.

Products for general laboratory use are not *in vitro* diagnostic medical devices unless such products, in view of their characteristics, are specifically intended by their manufacturer to be used for *in vitro* diagnostic examination;

iv. 'custom-made device' means any device specifically made in accordance with a duly qualified medical practitioner's written prescription which gives, under his responsibility, specific design characteristics and is intended for the sole use of a particular patient.

The abovementioned prescription may also be made out by any other person authorized by virtue of his professional qualifications to do so.

Mass-produced devices which need to be adapted to meet the specific requirements of the medical practitioner or any other professional user shall not be considered to be custom-made devices;

v. 'device intended for clinical investigation' means any device intended for use by a duly qualified medical practitioner when conducting investigations as referred to in Section 2.1 of Annex X in an adequate human clinical environment.

For the purpose of conducting clinical investigation, any other person who, by virtue of his professional qualifications, is authorized to carry out such investigation shall be accepted as equivalent to a duly qualified medical practitioner;

vi. 'manufacturer' means the natural or legal person with responsibility for the design, manufacture, packaging, and labeling of a device before it is placed on the market under his own name, regardless of whether these operations are carried out by that person himself or on his behalf by a third party.

The obligations of this Directive to be met by manufacturers also apply to the natural or legal person who assembles, packages, processes, fully refurbishes and/or labels one or more ready-made products and/or assigns to them their intended purpose as a device with a view to their being placed on the market under his own name. This subparagraph does not apply to the person who, while not a manufacturer within the meaning of the first subparagraph, assembles or adapts devices already on the market to their intended purpose for an individual patient;

vii. 'intended purpose' means the use for which the device is intended according to the data supplied by the manufacturer on the labeling, in the instructions and/or in promotional materials;

viii. 'placing on the market' means the first making available in return for payment or free of charge of a device other than a device intended for clinical investigation, with a view to distribution and/or use on the Community market, regardless of whether it is new or fully refurbished;

ix. 'putting into service' means the stage at which a device has been made available to the final user as being ready for use on the Community market for the first time for its intended purpose;

x. 'authorized representative' means any natural or legal person established in the Community who, explicitly designated by the manufacturer, acts, and maybe addressed by authorities and bodies in the Community instead of the manufacturer with regard to the latter's obligations under this Directive;

xi. 'clinical data' means the safety and/or performance information that is generated from the use of a device. Clinical data are sourced from:

a. clinical investigation(s) of the device concerned; or

b. clinical investigation(s) or other studies reported in the scientific literature, of a similar device for which equivalence to the device in question can be demonstrated; or

c. published and/or unpublished reports on other clinical experience of either the device in question or a similar device for which equivalence to the device in question can be demonstrated;

xii. 'device subcategory' means a set of devices having common areas of intended use or common technology;

xiii. 'generic device group' means a set of devices having the same or similar intended uses or commonality of technology allowing them to be classified in a generic manner not reflecting specific characteristics;

xiv. 'single-use device' means a device intended to be used once only for a single patient.

3. Where a device is intended to administer a medicinal product within the meaning of Article 1 of Directive 2001/83/EC (1), that device shall be governed by this Directive, without prejudice to the provisions of Directive 2001/83/EC with regard to the medicinal product.

If, however, such a device is placed on the market in such a way that the device and the medicinal product form a single integral product which is intended exclusively for use in the given combination and which is not reusable, that single product shall be governed by Directive 2001/83/EC. The relevant essential requirements of Annex I to this Directive shall apply as far as safety and performance-related device features are concerned.

4. Where a device incorporates, as an integral part, a substance which, if used separately, may be considered to be a medicinal product within the meaning of Article 1 of Directive 2001/83/EC and which is liable to act upon the body with action ancillary to that of the device, that device shall be assessed and authorized in accordance with this directive.

4a. Where a device incorporates, as an integral part, a substance which, if used separately, may be considered to be a medicinal product constituent or a medicinal product derived from human blood or human plasma within the meaning of Article 1 of Directive 2001/83/EC and which is liable to act upon the human body with action ancillary to that of the device, hereinafter referred to as a 'human blood derivative,' that device shall be assessed and authorized in accordance with this Directive.

5. This Directive shall not apply to:

i. *in vitro* diagnostic devices;

ii. active implantable devices covered by Directive 90/385/EEC;

iii. medicinal products covered by Directive 2001/83/EC. In deciding whether a product falls under that Directive or this Directive, particular account shall be taken of the principal mode of action of the product;

iv. cosmetic products covered by Directive 76/768/EEC (1); e) human blood, blood products, plasma or blood cells of human origin or to devices which incorporate at the time of placing on the market such blood products, plasma or cells, with the exception of devices referred to in paragraph 4a;

v. transplants or tissues or cells of human origin nor to products incorporating or derived from tissues or cells of human origin, with the exception of devices referred to in paragraph 4a;

vi. transplants or tissues or cells of animal origin, unless a device is manufactured utilizing animal tissue which is rendered non-viable or non-viable products derived from animal tissue.

6. Where a device is intended by the manufacturer to be used in accordance with both the provisions on personal protective equipment in Council Directive 89/686/EEC (1) and this directive, the relevant basic health and safety requirements of Directive 89/686/EEC shall also be fulfilled.

7. This directive is a specific directive within the meaning of Article 1(4) of Directive 2004/108/EC of the European Parliament and of the Council (2).

8. This Directive shall not affect the application of Council Directive 96/29/Euratom of 13 May 1996 laying down basic safety standards for the protection of the health of workers and the general public against the dangers arising from ionizing radiation (3), nor of Council Directive 97/43/Euratom of 30 June 1997 on health protection of individuals against the dangers of ionizing radiation in relation to medical exposure.

The Medical Devices Directive outlines the minimum requirements a manufacturer should meet when aiming to ensure the safety and performance characteristics for Medical Devices in the European market. It covers the following principles:

- The Essential Requirements: this part of the directive outlines the necessary precautions and requirements to be considered in the design, manufacture, use, and disposal of medical devices.
- Classification of Devices and Conformity Assessment Procedures which are based on the level of risk inherent in the device.
- Control of Production to ensure conformity to type.
- monitoring and vigilance.

Manufacturers of Medical Devices should demonstrate conformity to all requirements listed within the directive in addition to the requirements in subsequent guideline documents such as MEDDEVs (medical device directives).

For many of the common techniques and product types, there exist what is known as harmonized standards. While the use of a harmonized standard is not always mandatory, it is highly recommended as it represents best practice and technical state of the art, which can be used as a presumption of conformity to the relevant part of the directive globally. A manufacturer may however choose to not follow a harmonized standard, which is also okay but, in this case, they must be able to demonstrate clearly how the product meets the relevant safety or performance requirements.

Article 5 of the MDD 93/42/EEC: Reference to Standards

1. Member States shall presume compliance with the essential requirements referred to in Article 3 in respect of devices which are in conformity with the relevant national standards adopted pursuant to the harmonized standards the references of which have been publishes in the Official Journal of the European Communities; Member States shall publish the references of such national standards.
2. For the purposes of this Directive, reference to harmonized standards also includes the monographs of the European Pharmacopoeia notably on surgical sutures and on interaction between medicinal products and materials used in devices containing such medicinal products, the references of which have been published in the Official Journal of the European Communities.
3. If a Member State or the Commission considers that the harmonized standards do not entirely meet the essential requirements referred to in Article 3, the measures to be taken by the Member States with regard to these standards and the publication referred to in paragraph 1 of this Article shall be adopted by the procedure defined in Article 6(2).

8.3. PRODUCTS NOT COVERED BY THE MDD

The directive does not apply to the following:

- *in vitro* diagnostic devices;
- active implantable devices covered by Directive 90/385/EEC;
- medicinal products covered by Directive 65/65/EEC, including medicinal products derived from blood as covered by Directive 89/381/EEC;
- cosmetic products covered by Directive 76/768/EEC.
- human blood, blood products, plasma or blood cells of human origin or to devices which incorporate at the time of placing on the market such blood products, plasma or cells, with the exception of devices referred to in paragraph 4a;
- transplants or tissues or cells of human origin nor to products incorporating or derived from tissues or cells of human origin;
- transplants or tissues or cells of animal origin, unless a device is manufactured utilizing animal tissue which is rendered non-viable or non-viable products derived from animal tissue.

Chapter

9

The Medical Device Regulation 2017/745

CONTENTS

The medical device regulation (MDR) was published on May 26, 2017, with the aim of replacing the medical device directive (MDD) and the Active Implantable Medical Devices Directive. The MDR will replace the current EU Medical Device Directive (93/42/EEC) and the EU's Directive on active implantable medical devices (AIMD) (90/385/EEC). All currently certified medical devices and AIMD must be recertified such that they comply with the new requirements of the MDR. The MDR was meant to come into effect on May 26, 2020, but this has now been delayed for a year to May 2021. The *in-vitro* diagnostic regulation (IVDR) replaced the IVD Directive, and comes into effect on May 26, 2022. The new MDR provides an additional time after the date of application allowing to place new products under the MDD for a grace period.

9.1. SCOPE OF THE MDR

The MDR clearly defines which types of devices will have to comply with the new requirements. The MDR now covers devices that previously fell under two separate European directives-the MDD and the Active Implantable Medical Devices Directive (AIMDD). The scope of the regulation is very clearly laid out in Article 1, as follows:

Chapter I: Scope and Definitions
Article 1: Subject Matter and Scope
1. This Regulation lays down rules concerning the placing on the market, making available on the market or putting into service of medical devices for human use and accessories for such devices in the Union. This Regulation also applies to clinical investigations concerning such medical devices and accessories conducted in the Union. 2. This Regulation shall also apply, as from the date of application of common specifications adopted pursuant to Article 9, to the groups of products without an intended medical purpose that are listed in Annex XVI, taking into account state of the art, and in particular existing harmonized standards for analogous devices with a medical purpose, based on similar technology. The common specifications for each of the groups of products listed in Annex XVI shall address, at least, application of risk management as set out in Annex I for the group of products in question and, where necessary, clinical evaluation regarding safety. The necessary common specifications shall be adopted by 26 May 2020. They shall apply as from six months after the date of their entry into force or from 26 May 2020, whichever is the latest. Notwithstanding Article 122, Member States' measures regarding the qualification of the products covered by Annex XVI as medical devices pursuant to Directive 93/42/EEC shall remain valid until the date of application, as referred to in the first subparagraph, of the relevant common specifications for that group of products.

This Regulation also applies to clinical investigations conducted in the Union concerning the products referred to in the first subparagraph.

3. Devices with both a medical and a non-medical intended purpose shall fulfill cumulatively the requirements applicable to devices with an intended medical purpose and those applicable to devices without an intended medical purpose.

4. For the purposes of this Regulation, medical devices, accessories for medical devices, and products listed in Annex XVI to which this Regulation applies pursuant to paragraph 2 shall hereinafter be referred to as 'devices.'

5. Where justified on account of the similarity between a device with an intended medical purpose placed on the market and a product without an intended medical purpose in respect of their characteristics and risks, the Commission is empowered to adopt delegated acts in accordance with Article 115 to amend the list in Annex XVI, by adding new groups of products, in order to protect the health and safety of users or other persons or other aspects of public health.

6. This Regulation does not apply to:

i. *in vitro* diagnostic medical devices covered by Regulation (EU) 2017/746;

ii. medicinal products as defined in point 2 of Article 1 of Directive 2001/83/EC. In deciding whether a product falls under Directive 2001/83/EC or under this Regulation, particular account shall be taken of the principal mode of action of the product;

iii. advanced therapy medicinal products covered by Regulation (EC) No 1394/2007;

iv. human blood, blood products, plasma or blood cells of human origin or devices which incorporate, when placed on the market or put into service, such blood products, plasma or cells, except for devices referred to in paragraph 8 of this Article;

v. cosmetic products covered by Regulation (EC) No 1223/2009;

vi. transplants, tissues or cells of animal origin, or their derivatives, or products containing or consisting of them; however, this Regulation does apply to devices manufactured utilizing tissues or cells of animal origin, or their derivatives, which are non-viable or are rendered non-viable;

vii. transplants, tissues or cells of human origin, or their derivatives, covered by Directive 2004/23/EC, or products containing or consisting of them; however, this Regulation does apply to devices manufactured utilizing derivatives of tissues or cells of human origin which are non-viable or are rendered non-viable;

viii. products, other than those referred to in points (d), (f) and (g), that contain or consist of viable biological material or viable organisms, including living micro-organisms, bacteria, fungi or viruses in order to achieve or support the intended purpose of the product;

ix. food covered by Regulation (EC) No 178/2002.

7. Any device which, when placed on the market or put into service, incorporates as an integral part an *in vitro* diagnostic medical device as defined in point 2 of Article 2 of Regulation (EU) 2017/746, shall be governed by this Regulation. The requirements of Regulation (EU) 2017/746 shall apply to the *in vitro* diagnostic medical device part of the device.

8. Any device which, when placed on the market or put into service, incorporates, as an integral part, a substance which, if used separately, would be considered to be a medicinal product as defined in point 2 of Article 1 of Directive 2001/83/EC, including a medicinal product derived from human blood or human plasma as defined in point 10 of Article 1 of that Directive, and that has an action ancillary to that of the device, shall be assessed and authorized in accordance with this Regulation.

However, if the action of that substance is principal and not ancillary to that of the device, the integral product shall be governed by Directive 2001/83/EC or Regulation (EC) No 726/2004 of the European Parliament and of the Council (1), as applicable. In that case, the relevant general safety and performance requirements set out in Annex I to this Regulation shall apply as far as the safety and performance of the device part are concerned.

9. Any device which is intended to administer a medicinal product as defined in point 2 of Article 1 of Directive 2001/83/EC shall be governed by this Regulation, without prejudice to the provisions of that Directive and of Regulation (EC) No 726/2004 with regard to the medicinal product.

However, if the device intended to administer a medicinal product and the medicinal product are placed on the market in such a way that they form a single integral product which is intended exclusively for use in the given combination and which is not reusable, that single integral product shall be governed by Directive 2001/83/EC or Regulation (EC) No 726/2004, as applicable. In that case, the relevant general safety and performance requirements set out in Annex I to this Regulation shall apply as far as the safety and performance of the device part of the single integral product are concerned.

10. Any device which, when placed on the market or put into service, incorporates, as an integral part, non-viable tissues or cells of human origin or their derivatives that have an action ancillary to that of the device shall be assessed and authorized in accordance with this Regulation. In that case, the provisions for donation, procurement, and testing laid down in Directive 2004/23/EC shall apply.

However, if the action of those tissues or cells or their derivatives is principal and not ancillary to that of the device and the product is not governed by Regulation (EC) No 1394/2007, the product shall be governed by Directive 2004/23/EC. In that case, the relevant general safety and performance requirements set out in Annex I to this Regulation shall apply as far as the safety and performance of the device part are concerned.

11. This Regulation is specific Union legislation within the meaning of Article 2(3) of Directive 2014/30/EU.

12. Devices that are also machinery within the meaning of point (a) of the second paragraph of Article 2 of Directive 2006/42/EC of the European Parliament and of the Council (1) shall, where a hazard relevant under that Directive exists, also meet the essential health and safety requirements set out in Annex I to that Directive to the extent to which those requirements are more specific than the general safety and performance requirements set out in Chapter II of Annex I to this Regulation.

13. This Regulation shall not affect the application of Directive 2013/59/Euratom.

14.	This Regulation shall not affect the right of a Member State to restrict the use of any specific type of device in relation to aspects not covered by this Regulation.
15.	This Regulation shall not affect national law concerning the organization, delivery or financing of health services and medical care, such as the requirement that certain devices may only be supplied on a medical prescription, the requirement that only certain health professionals or healthcare institutions may dispense or use certain devices or that their use be accompanied by specific professional counseling.
16.	Nothing in this Regulation shall restrict the freedom of the press or the freedom of expression in the media in so far as those freedoms are guaranteed in the Union and in the Member States, in particular under Article 11 of the Charter of Fundamental Rights of the European Union.

The MDR defines a medical device as any "instrument, apparatus, appliance, software, implant, reagent, material or other article" intended to be used for any of the following medical purposes:

- Diagnosis, prevention, monitoring, treatment or alleviation of disease, disability or injury, where prevention of disability and injury is excluded.
- Investigation, replacement or modification of an anatomical, physiological or pathological process.
- Providing data via *in vitro* examination of samples derived from a human body.

The basic definitions as defined in the MDR are given in Article 2 as shown in the box below:

Article 2: Definitions

For the purposes of this Regulation, the following definitions apply:

1. 'Medical device' means any instrument, apparatus, appliance, software, implant, reagent, material or other article intended by the manufacturer to be used, alone or in combination, for human beings for one or more of the following specific medical purposes:

i. diagnosis, prevention, monitoring, prediction, prognosis, treatment or alleviation of disease.

ii. diagnosis, monitoring, treatment, alleviation of, or compensation for, an injury or disability.

iii. investigation, replacement or modification of the anatomy or of a physiological or pathological process or state.

iv. providing information by means of *in vitro* examination of specimens derived from the human body, including organ, blood, and tissue donations, and which does not achieve its principal intended action by pharmacological, immunological or metabolic means, in or on the human body, but which may be assisted in its function by such means.

The following products shall also be deemed to be medical devices:

a. devices for the control or support of conception;

b. products specifically intended for the cleaning, disinfection or sterilization of devices as referred to in Article 1(4) and of those referred to in the first paragraph of this point.

2. 'Accessory for a medical device' means an article which, whilst not being itself a medical device, is intended by its manufacturer to be used together with one or several particular medical device(s) to specifically enable the medical device(s) to be used in accordance with its/their intended purpose(s) or to specifically and directly assist the medical functionality of the medical device(s) in terms of its/their intended purpose(s);

3. 'Custom-made device' means any device specifically made in accordance with a written prescription of any person authorized by national law by virtue of that person's professional qualifications which gives, under that person's responsibility, specific design characteristics, and is intended for the sole use of a particular patient exclusively to meet their individual conditions and needs.

However, mass-produced devices which need to be adapted to meet the specific requirements of any professional user and devices which are mass-produced by means of industrial manufacturing processes in accordance with the written prescriptions of any authorized person shall not be considered to be custom-made devices;

4. 'Active device' means any device, the operation of which depends on a source of energy other than that generated by the human body for that purpose, or by gravity, and which acts by changing the density of or converting that energy. Devices intended to transmit energy, substances or other elements between an active device and the patient, without any significant change, shall not be deemed to be active devices.

Software shall also be deemed to be an active device;

5. 'Implantable device' means any device, including those that are partially or wholly absorbed, which is intended:

i. to be totally introduced into the human body, or

ii. to replace an epithelial surface or the surface of the eye, by clinical intervention and which is intended to remain in place after the procedure.

Any device intended to be partially introduced into the human body by clinical intervention and intended to remain in place after the procedure for at least 30 days shall also be deemed to be an implantable device;

6. 'Invasive device' means any device which, in whole or in part, penetrates inside the body, either through a body orifice or through the surface of the body;

7. 'Generic device group' means a set of devices having the same or similar intended purposes or a commonality of technology allowing them to be classified in a generic manner not reflecting specific characteristics;

8. 'Single-use device' means a device that is intended to be used on one individual during a single procedure;

9. 'Falsified device' means any device with a false presentation of its identity and/or of its source and/or its CE marking certificates or documents relating to CE marking procedures. This definition does not include unintentional non-compliance and is without prejudice to infringements of intellectual property rights;

10. 'Procedure pack' means a combination of products packaged together and placed on the market with the purpose of being used for a specific medical purpose;

11. 'System' means a combination of products, either packaged together or not, which are intended to be interconnected or combined to achieve a specific medical purpose;

12. 'Intended purpose' means the use for which a device is intended according to the data supplied by the manufacturer on the label, in the instructions for use or in promotional or sales materials or statements and as specified by the manufacturer in the clinical evaluation;

13. 'Label' means the written, printed or graphic information appearing either on the device itself, or on the packaging of each unit or on the packaging of multiple devices;

14. 'Instructions for use' means the information provided by the manufacturer to inform the user of a device's intended purpose and proper use and of any precautions to be taken;

15. 'Unique device identifier (UDI)' means a series of numeric or alphanumeric characters that is created through internationally accepted device identification and coding standards and that allows unambiguous identification of specific devices on the market;

16. 'Non-viable' means having no potential for metabolism or multiplication;

17. 'Derivative' means a 'non-cellular substance' extracted from human or animal tissue or cells through a manufacturing process. The final substance used for manufacturing of the device in this case does not contain any cells or tissues;

18.	'Nanomaterial' means a natural, incidental or manufactured material containing particles in an unbound state or as an aggregate or as an agglomerate and where, for 50% or more of the particles in the number size distribution, one or more external dimensions is in the size range 1–100 nm; Fullerenes, graphene flakes and single-wall carbon nanotubes with one or more external dimensions below 1 nm shall also be deemed to be nanomaterials;
19.	'Particle,' for the purposes of the definition of nanomaterial in point (18), means a minute piece of matter with defined physical boundaries;
20.	'Agglomerate,' for the purposes of the definition of nanomaterial in point (18), means a collection of weakly bound particles or aggregates where the resulting external surface area is similar to the sum of the surface areas of the individual components;
21.	'Aggregate,' for the purposes of the definition of nanomaterial in point (18), means a particle comprising of strongly bound or fused particles;
22.	'Performance' means the ability of a device to achieve its intended purpose as stated by the manufacturer;
23.	'Risk' means the combination of the probability of occurrence of harm and the severity of that harm;
24.	'Benefit-risk determination' means the analysis of all assessments of benefit and risk of possible relevance for the use of the device for the intended purpose, when used in accordance with the intended purpose given by the manufacturer;
25.	'Compatibility' is the ability of a device, including software, when used together with one or more other devices in accordance with its intended purpose, to:
i.	Perform without losing or compromising the ability to perform as intended, and/or
ii.	Integrate and/or operate without the need for modification or adaption of any part of the combined devices, and/or
iii.	Be used together without conflict/interference or adverse reaction.
26.	'Interoperability' is the ability of two or more devices, including software, from the same manufacturer or from different manufacturers, to:
i.	Exchange information and use the information that has been exchanged for the correct execution of a specified function without changing the content of the data, and/or
ii.	Communicate with each other, and/or
iii.	Work together as intended.
27.	'Making available on the market' means any supply of a device, other than an investigational device, for distribution, consumption or use on the Union market in the course of a commercial activity, whether in return for payment or free of charge;
28.	'Placing on the market' means the first making available of a device, other than an investigational device, on the Union market;
29.	'Putting into service' means the stage at which a device, other than an investigational device, has been made available to the final user as being ready for use on the Union market for the first time for its intended purpose;

30. 'Manufacturer' means a natural or legal person who manufactures or fully refurbishes a device or has a device designed, manufactured or fully refurbished, and markets that device under its name or trademark;
31. 'Fully refurbishing,' for the purposes of the definition of manufacturer, means the complete rebuilding of a device already placed on the market or put into service, or the making of a new device from used devices, to bring it into conformity with this Regulation, combined with the assignment of a new lifetime to the refurbished device;
32. 'Authorized representative' means any natural or legal person established within the Union who has received and accepted a written mandate from a manufacturer, located outside the Union, to act on the manufacturer's behalf in relation to specified tasks with regard to the latter's obligations under this Regulation;
33. 'Importer' means any natural or legal person established within the Union that places a device from a third country on the Union market;
34. 'Distributor' means any natural or legal person in the supply chain, other than the manufacturer or the importer, that makes a device available on the market, up until the point of putting into service;
35. 'Economic operator' means a manufacturer, an authorized representative, an importer, a distributor or the person referred to in Article 22(1) and 22(3);
36. 'Health institution' means an organization the primary purpose of which is the care or treatment of patients or the promotion of public health;
37. 'User' means any healthcare professional or layperson who uses a device;
38. 'Lay person' means an individual who does not have formal education in a relevant field of healthcare or medical discipline;
39. 'Reprocessing' means a process carried out on a used device in order to allow its safe reuse including cleaning, disinfection, sterilization, and related procedures, as well as testing and restoring the technical and functional safety of the used device;
40. 'Conformity assessment' means the process demonstrating whether the requirements of this Regulation relating to a device have been fulfilled;
41. 'Conformity assessment body' means a body that performs third-party conformity assessment activities including calibration, testing, certification, and inspection;
42. 'Notified body' means a conformity assessment body designated in accordance with this Regulation;
43. 'CE marking of conformity' or 'CE marking' means a marking by which a manufacturer indicates that a device is in conformity with the applicable requirements set out in this Regulation and other applicable Union harmonization legislation providing for its affixing;
44. 'Clinical evaluation' means a systematic and planned process to continuously generate, collect, analyze, and assess the clinical data pertaining to a device in order to verify the safety and performance, including clinical benefits, of the device when used as intended by the manufacturer;
45. 'Clinical investigation' means any systematic investigation involving one or more human subjects, undertaken to assess the safety or performance of a device;
46. 'Investigational device' means a device that is assessed in a clinical investigation;

47. 'Clinical investigation plan' means a document that describes the rationale, objectives, design, methodology, monitoring, statistical considerations, organization, and conduct of a clinical investigation;

48. 'Clinical data' means information concerning safety or performance that is generated from the use of a device and is sourced from the following:

i. clinical investigation(s) of the device concerned.

ii. clinical investigation(s) or other studies reported in scientific literature, of a device for which equivalence to the device in question can be demonstrated.

iii. reports published in peer reviewed scientific literature on other clinical experience of either the device in question or a device for which equivalence to the device in question can be demonstrated.

iv. clinically relevant information coming from post-market surveillance, in particular the post-market clinical follow-up;

49. 'Sponsor' means any individual, company, institution or organization which takes responsibility for the initiation, for the management and setting up of the financing of the clinical investigation;

50. 'Subject' means an individual who participates in a clinical investigation;

51. 'Clinical evidence' means clinical data and clinical evaluation results pertaining to a device of a sufficient amount and quality to allow a qualified assessment of whether the device is safe and achieves the intended clinical benefit(s), when used as intended by the manufacturer;

52. 'Clinical performance' means the ability of a device, resulting from any direct or indirect medical effects which stem from its technical or functional characteristics, including diagnostic characteristics, to achieve its intended purpose as claimed by the manufacturer, thereby leading to a clinical benefit for patients, when used as intended by the manufacturer;

53. 'Clinical benefit' means the positive impact of a device on the health of an individual, expressed in terms of a meaningful, measurable, patient-relevant clinical outcome(s), including outcome(s) related to diagnosis, or a positive impact on patient management or public health;

54. 'Investigator' means an individual responsible for the conduct of a clinical investigation at a clinical investigation site;

55. 'Informed consent' means a subject's free and voluntary expression of his or her willingness to participate in a particular clinical investigation, after having been informed of all aspects of the clinical investigation that are relevant to the subject's decision to participate or, in the case of minors and of incapacitated subjects, an authorization or agreement from their legally designated representative to include them in the clinical investigation;

56. 'Ethics committee' means an independent body established in a Member State in accordance with the law of that Member State and empowered to give opinions for the purposes of this Regulation, taking into account the views of laypersons, in particular patients or patients' organizations;

57. 'Adverse event' means any untoward medical occurrence, unintended disease or injury or any untoward clinical signs, including an abnormal laboratory finding, in subjects, users or other persons, in the context of a clinical investigation, whether or not related to the investigational device;

58. 'Serious adverse event' means any adverse event that led to any of the following:

i. death.

ii. serious deterioration in the health of the subject, that resulted in any of the following:

a. life-threatening illness or injury.

b. permanent impairment of a body structure or a body function.

c. hospitalization or prolongation of patient hospitalization.

d. medical or surgical intervention to prevent life-threatening illness or injury or permanent impairment to a body structure or a body function.

e. chronic disease.

iii. fetal distress, fetal death or a congenital physical or mental impairment or birth defect;

59. 'Device deficiency' means any inadequacy in the identity, quality, durability, reliability, safety or performance of an investigational device, including malfunction, use errors or inadequacy in information supplied by the manufacturer;

60. 'Post-market surveillance' means all activities carried out by manufacturers in cooperation with other economic operators to institute and keep up to date a systematic procedure to proactively collect and review experience gained from devices they place on the market, make available on the market or put into service for the purpose of identifying any need to immediately apply any necessary corrective or preventive actions;

61. 'Market surveillance' means the activities carried out and measures taken by competent authorities to check and ensure that devices comply with the requirements set out in the relevant Union harmonization legislation and do not endanger health, safety or any other aspect of public interest protection;

62. 'Recall' means any measure aimed at achieving the return of a device that has already been made available to the end-user;

63. 'Withdrawal' means any measure aimed at preventing a device in the supply chain from being further made available on the market;

64. 'Incident' means any malfunction or deterioration in the characteristics or performance of a device made available on the market, including use-error due to ergonomic features, as well as any inadequacy in the information supplied by the manufacturer and any undesirable side-effect;

65. 'Serious incident' means any incident that directly or indirectly led, might have led or might lead to any of the following:

i. the death of a patient, user or other person.

ii. the temporary or permanent serious deterioration of a patient's, user's or other person's state of health.

iii. a serious public health threat;

66. 'serious public health threat' means an event which could result in imminent risk of death, serious deterioration in a person's state of health, or serious illness, that may require prompt remedial action, and that may cause significant morbidity or mortality in humans, or that is unusual or unexpected for the given place and time;

67. 'Corrective action' means action taken to eliminate the cause of a potential or actual non-conformity or other undesirable situation;

68. 'Field safety corrective action' means corrective action taken by a manufacturer for technical or medical reasons to prevent or reduce the risk of a serious incident in relation to a device made available on the market;

69. 'Field safety notice' means a communication sent by a manufacturer to users or customers in relation to a field safety corrective action;

70. 'Harmonized standard' means a European standard as defined in point (1)(c) of Article 2 of Regulation (EU) No 1025/2012;

71. 'Common specifications' (CS) means a set of technical and/or clinical requirements, other than a standard, that provides a means of complying with the legal obligations applicable to a device, process, or system.

The new MDR also specifies particular types of products that qualify as medical devices, such as products intended for cleaning, disinfection, and sterilization of medical devices as well as devices for the control and support of conception. Manufacturers that are currently certified to market their devices in the European Union (EU) should also familiarize themselves with the MDR context so as to determine whether their existing products still fall within the scope of the new regulation and if the classification has changed. Some simplified examples are presented in Table 9.1.

Table 9.1: Examples of Potential Medical Devices Classified Under the New Medical Device Regulation 2017/745

Examples of Potential Medical Products	Are they Classified as a Medical Device Under the MDR
Products used for cleaning, disinfecting or sterilizing other medical devices (e.g., scope disinfector)	Yes
Products that utilize viable biological components for their intended purpose (e.g., lactobacillus)	No
Devices for purposes of control or support of conception (e.g., condoms)	Yes
Software used for any purpose covered by MDR's definition of a medical device	yes
IVDs, medicinal products, cosmetics, and food products; these are regulated elsewhere	No
Software used for general administrative purposes, even in healthcare settings	No

In addition to the definitions defined in Article 2, Annex XVI of the new regulation also lists groups of products that, despite having no intended medical purpose, must also meet the MDR requirements. The major groups of products defined in Annex XVI include:

- Contact lenses and other products used in or on the eye.
- Products introduced into the body via surgically invasive means in order to modify anatomy.
- Products and substances used for facial or other subcutaneous filling.
- Equipment used for liposuction, lipolysis or lipoplasty.
- High-intensity electromagnetic radiation equipment used for skin treatments such as tattoo and hair removal.
- Equipment using electrical or magnetic currents to stimulate the brain.

9.1.1. Accessories

Accessories are defined as products that are intended to enable a device to be used in accordance with its intended purpose or to assist the functionality of the device itself. While accessories are not medical devices per se, they are covered within the scope of the regulation and fall under the definition of a "device."

Annex XVI: List of Groups of Products Without an Intended Medical Purpose Referred to in Article 1(2)	
1.	Contact lenses or other items intended to be introduced into or onto the eye.
2.	Products intended to be totally or partially introduced into the human body through surgically invasive means for the purpose of modifying the anatomy or fixation of body parts with the exception of tattooing products and piercings.
3.	Substances, combinations of substances, or items intended to be used for facial or other dermal or mucous membrane filling by subcutaneous, submucous or intradermal injection or other introduction, excluding those for tattooing.
4.	Equipment intended to be used to reduce, remove or destroy adipose tissue, such as equipment for liposuction, lipolysis or lipoplasty.
5.	High intensity electromagnetic radiation (e.g., infra-red, visible light and ultra-violet) emitting equipment intended for use on the human body, including coherent and non-coherent sources, monochromatic, and broad-spectrum, such as lasers and intense pulsed light equipment, for skin resurfacing, tattoo or hair removal or other skin treatment.
6.	Equipment intended for brain stimulation that apply electrical currents or magnetic or electromagnetic fields that penetrate the cranium to modify neuronal activity in the brain.

9.2. MAJOR DIFFERENCES BETWEEN THE MEDICAL DEVICE DIRECTIVE 93/42/EEC AND MEDICAL DEVICE REGULATION (MDR) 2017/745

The basic structure and requirements of the Regulation remain the same as that of the medical device directive (MDD) however a summary of the major differences between the MDD 93/42/EEC and the MDR 2017/745 is provided below:

- There is a wider and more detailed scope of EU legislation which has been extended to include a greater variety of devices such as implants for esthetic purposes.
- There is much stronger supervision of notified bodies by national authorities.
- The rights and responsibilities for manufacturers, authorized representatives (AR), importers, and distributors have been defined more clearly.
- There is an extended database on medical devices (known as EUDAMED) which aims to provide a more comprehensive and public information source on devices that are available on the EU market.
- More stringent rules for clinical investigations on devices.
- Introduction of classification rules that divide the broad range of *in vitro* diagnostic (IVD) medical devices into four different risk classes as it already exists for other medical devices.
- Creation of a Medical Device Coordination Group (MDCG) composed of members representing national competent authorities in the field of medical devices to ensure better coordination between Member States.

Chapter

10 Economic Operators

CONTENTS

10.1. INTRODUCTION TO ECONOMIC OPERATORS

It is a common misconception that only the manufacturer is responsible for ensuring compliance of medical devices with the relevant regulation. in fact, it is the responsibility of the manufacturer as well as the distributor, importer, authorized representative and notified body to name a few; all of which are now responsible for ensuring that the medical devices reaching the market are fully compliant. Both the EU medical devices regulation (MDR) and *in vitro* diagnostic medical devices regulation (IVDR) contain a complete section on economic operators for medical devices and introduces new requirements. Although the manufacturer is ultimately responsible for ensuring their medical device is compliant with the relevant legislation, the various defined economic operators also have a part to play in ensuring the safety and performance of the device throughout the supply chain. This new policy of sharing compliance across the supply chain is an important component of the EU MDR with each economic operator serving as a control on the other, causing each device to receive a number of checks throughout the process before it is brought to market. In addition to this, it also allows for any compliance issues emerging from any one of the economic operators to have a direct legal implication on the other.

Since 2008, the EU has gradually implemented the requirements for economic operators in every CE marking directive and regulation that was renewed. Now, the new roles and responsibilities of these economic operators have also appeared in the MDR and IVDR which has further strengthened their quality and regulatory compliance responsibilities.

10.2. WHAT IS AN ECONOMIC OPERATOR?

The term "economic operator" refers to manufacturers, authorized representatives (AR), distributors, and importers. Under the new MDR and IVDR, these economic operators will share the duty of ensuring compliance of the medical devices they manufacture, import, and distribute within the EU. These four entities considered as Economic Operators are defined in Article 2 Definitions of the MDR. The definitions present within the MDR is presented below:

Article 2: Definitions
'Authorized representative' means any natural or legal person established within the Union who has received and accepted a written mandate from a manufacturer, located outside the Union, to act on the manufacturer's behalf in relation to specified tasks with regard to the latter's obligations under this Regulation; 'Importer' means any natural or legal person established within the Union that places a device from a third country on the Union market; 'Distributor' means any natural or legal person in the supply chain, other than the manufacturer or the importer, that makes a device available on the market, up until the point of putting into service; 'Economic operator' means a manufacturer, an authorized representative, an importer, a distributor or the person referred to in Article 22(1) and 22(3); 'Manufacturer' means a natural or legal person who manufactures or fully refurbishes a device or has a device designed, manufactured or fully refurbished, and markets that device under its name or trademark.

It's important to understand what each economic operator-defined under the MDR is responsible for. Manufacturers are responsible for EUDAMED registration, technical documentation, design, development, manufacture, and assembly, as well as handling, storage, distribution, corrective actions, UDI labeling, complaints, post-market surveillance and assigning a PRRC. AR are responsible for EUDAMED registration, technical documentation, corrective actions, UDI labeling, post-market surveillance and PRRC. Importers are responsible for EUDAMED registration, handling, storage, and distribution, corrective actions, UDI labeling, and post-market surveillance, while distributors are responsible for handling, storage, and distribution, corrective actions, UDI labeling, complaints, and post-market surveillance.

10.3. WHO CONTROLS ECONOMIC OPERATORS?

There is a great deal of confusion when it comes to Economic Operators located within and outside of the EU. Often EO's are unsure if they should contact a Notified Body to sell medical devices in the EU. Often EO's don't contact a NB however they can still receive a visit from a NB. The only entity that should be contacting the NB, other than the manufacturer is the Competent Authority (CA). For distributors and/or importers, the CA of the country in which the devices are being sold in will have the responsibility to check that the respective EO is doing the right job; however, the frequency with which they may choose to inspect the EO is not defined. This is in contrast to a NB which will audit the EO every year to ensure performance as required by the regulation.

For example, if a distributor is not complying with the storage conditions of the products which they are distributing or if the final users have raised some issues or adverse events with the final product, then a CA may choose to inspect the respective EO. AR can also be inspected by a CA as well as a NB if the AR's activities are in the scope of the QMS.

10.4. DO ECONOMIC OPERATORS NEED A QUALITY MANAGEMENT SYSTEM (QMS)?

This question often leads to some confusion amongst Economic Operators as the requirement for a QMS is not clearly mandated unless you are a manufacturer (refer to Article 10 of the MDR). Ideally all EO's should at least still organize their processes and flow of information in a similar way to that of a suitably implemented QMS.

The onus is often on the manufacturer to decide if working with an EO that has no QMS implemented will still be beneficial to the overall distribution and importation of safe medical devices. It is still important to note that while the MDR does not make mention that EO's need to have a QMS implemented, there are still many procedural requirements which these EO's have to comply with. These are often outlined in guidance documents for good distribution practices (GDP) so as to ensure the safe and controlled delivery of medical devices to the market.

Within the EU MDR and EU IVRD, Economic Operators have clearly defined responsibilities which have been defined in relevant articles as named below:

- **Article 10:** Manufacturers.
- **Article 11:** Authorized Representatives.
- **Article 13:** General Obligations of Importers.
- **Article 14:** General Obligations of Distributors.

These obligations are further discussed in the chapters below.

10.5. GENERAL OBLIGATIONS OF MANUFACTURERS AS ECONOMIC OPERATORS

The new MDR also defines the main obligations of manufacturers. This includes in summary:

- that devices designed and manufactured are done so in accordance with the requirements of the regulation itself.
- the ability to establish, document, implement, and maintain a system for risk management.
- ensuring a clinical evaluation is conducted.
- drawing up suitable technical documentation and keeping them up to date.
- drawing up an EU declaration of conformity (DoC) and affixing a CE mark.
- ensuring retention of documentation such shall be available for competent authorities for at least 10 years and 15 years for implantable devices.
- Ensuring that procedures that ensure production is kept in conformity with the relevant legal requirements.
- Implementing a post-market surveillance system and ensuring it is kept up to date.
- Ensure that the device is accompanied by the appropriate information.
- Taking the necessary corrective actions when the manufacturer has reason to believe that a device they have placed onto the European market or put into service is not in conformity with the legal requirements.
- Implement a system for the recording and reporting of incidents and field safety corrective actions (FSCA).
- Have measures in place regarding financial coverage for potential liability as natural or legal persons may claim compensation for damage caused by a defective device.

The general obligations of manufacturers as per the MDR is provided in the box below:

Article 10: General Obligations of Manufacturers
1. When placing their devices on the market or putting them into service, manufacturers shall ensure that they have been designed and manufactured in accordance with the requirements of this Regulation. 2. Manufacturers shall establish, document, implement, and maintain a system for risk management as described in Section 3 of Annex I. 3. Manufacturers shall conduct a clinical evaluation in accordance with the requirements set out in Article 61 and Annex XIV, including a PMCF. 4. Manufacturers of devices other than custom-made devices shall draw up and keep up-to-date technical documentation for those devices. The technical documentation shall be such as to allow the conformity of the device with the requirements of this Regulation to be assessed. The technical documentation shall include the elements set out in Annexes II and III. The Commission is empowered to adopt delegated acts in accordance with Article 115 amending, in the light of technical progress, the Annexes II and III. 5. Manufacturers of custom-made devices shall draw up, keep up to date and keep available for competent authorities documentation in accordance with Section 2 of Annex XIII. 6. Where compliance with the applicable requirements has been demonstrated following the applicable conformity assessment procedure, manufacturers of devices, other than custom-made or investigational devices, shall draw up an EU declaration of conformity in accordance with Article 19, and affix the CE marking of conformity in accordance with Article 20. 7. Manufacturers shall comply with the obligations relating to the UDI system referred to in Article 27 and with the registration obligations referred to in Articles 29 and 31. 8. Manufacturers shall keep the technical documentation, the EU declaration of conformity and, if applicable, a copy of any relevant certificate, including any amendments and supplements, issued in accordance with Article 56, available for the competent authorities for a period of at least 10 years after the last device covered by the EU declaration of conformity has been placed on the market. In the case of implantable devices, the period shall be at least 15 years after the last device has been placed on the market. Upon request by a competent authority, the manufacturer shall, as indicated therein, provide that technical documentation in its entirety or a summary thereof. A manufacturer with a registered place of business outside the Union shall, in order to allow its authorized representative to fulfill the tasks mentioned in Article 11(3), ensure that the authorized representative has the necessary documentation permanently available. 9. Manufacturers shall ensure that procedures are in place to keep series production in conformity with the requirements of this Regulation. Changes in device design or characteristics and changes in the harmonized standards or CS by reference to which the conformity of a device is declared shall be adequately taken into account in a timely manner.

Manufacturers of devices, other than investigational devices, shall establish, document, implement, maintain, keep up to date and continually improve a quality management system that shall ensure compliance with this Regulation in the most effective manner and in a manner that is proportionate to the risk class and the type of device.

The quality management system shall cover all parts and elements of a manufacturer's organization dealing with the quality of processes, procedures, and devices. It shall govern the structure, responsibilities, procedures, processes, and management resources required to implement the principles and actions necessary to achieve compliance with the provisions of this Regulation.

The quality management system shall address at least the following aspects:

i. a strategy for regulatory compliance, including compliance with conformity assessment procedures and procedures for management of modifications to the devices covered by the system;

ii. identification of applicable general safety and performance requirements and exploration of options to address those requirements;

iii. responsibility of the management;

iv. resource management, including selection and control of suppliers and sub-contractors;

v. risk management as set out in Section 3 of Annex I;

vi. clinical evaluation in accordance with Article 61 and Annex XIV, including PMCF;

vii. product realization, including planning, design, development, production, and service provision;

viii. verification of the UDI assignments made in accordance with Article 27(3) to all relevant devices and ensuring consistency and validity of information provided in accordance with Article 29;

ix. setting-up, implementation, and maintenance of a post-market surveillance system, in accordance with Article 83;

x. handling communication with competent authorities, notified bodies, other economic operators, customers, and/or other stakeholders;

xi. processes for reporting of serious incidents and field safety corrective actions in the context of vigilance;

xii. management of corrective and preventive actions and verification of their effectiveness;

xiii. processes for monitoring and measurement of output, data analysis and product improvement.

10. Manufacturers of devices shall implement and keep up to date the post-market surveillance system in accordance with Article 83.

11. Manufacturers shall ensure that the device is accompanied by the information set out in Section 23 of Annex I in an official Union language(s) determined by the Member State in which the device is made available to the user or patient. The particulars on the label shall be indelible, easily legible and clearly comprehensible to the intended user or patient.

12. Manufacturers who consider or have reason to believe that a device which they have placed on the market or put into service is not in conformity with this Regulation shall immediately take the necessary corrective action to bring that device into conformity, to withdraw it or to recall it, as appropriate. They shall inform the distributors of the device in question and, where applicable, the authorized representative and importers accordingly.

Where the device presents a serious risk, manufacturers shall immediately inform the competent authorities of the Member States in which they made the device available and, where applicable, the notified body that issued a certificate for the device in accordance with Article 56, in particular, of the non-compliance and of any corrective action taken.

13. Manufacturers shall have a system for recording and reporting of incidents and field safety corrective actions as described in Articles 87 and 88.

14. Manufacturers shall, upon request by a competent authority, provide it with all the information and documentation necessary to demonstrate the conformity of the device, in an official Union language determined by the Member State concerned. The competent authority of the Member State in which the manufacturer has its registered place of business may require that the manufacturer provide samples of the device free of charge or, where that is impracticable, grant access to the device. Manufacturers shall cooperate with a competent authority, at its request, on any corrective action taken to eliminate or, if that is not possible, mitigate the risks posed by devices which they have placed on the market or put into service.

If the manufacturer fails to cooperate or the information and documentation provided is incomplete or incorrect, the competent authority may, in order to ensure the protection of public health and patient safety, take all appropriate measures to prohibit or restrict the device's being made available on its national market, to withdraw the device from that market or to recall it until the manufacturer cooperates or provides complete and correct information.

If a competent authority considers or has reason to believe that a device has caused damage, it shall, upon request, facilitate the provision of the information and documentation referred to in the first subparagraph to the potentially injured patient or user and, as appropriate, the patient's or user's successor in title, the patient's or user's health insurance company or other third parties affected by the damage caused to the patient or user, without prejudice to data protection rules and, unless there is an overriding public interest in disclosure, without prejudice to the protection of intellectual property rights.

The competent authority need not comply with the obligation laid down in the third subparagraph where disclosure of the information and documentation referred to in the first subparagraph is ordinarily dealt with in the context of legal proceedings.

15. Where manufacturers have their devices designed or manufactured by another legal or natural person, the information on the identity of that person shall be part of the information to be submitted in accordance with Article 30(1).

16. Natural or legal persons may claim compensation for damage caused by a defective device in accordance with applicable Union and national law.

> Manufacturers shall, in a manner that is proportionate to the risk class, type of device and the size of the enterprise, have measures in place to provide sufficient financial coverage in respect of their potential liability under Directive 85/374/EEC, without prejudice to more protective measures under national law.

10.6. GENERAL OBLIGATIONS OF IMPORTERS AS ECONOMIC OPERATORS

In some instances, companies may serve as an importer of a medical device. If this is the case, the company must still ensure it is in conformity with the regulation. Importers must ensure that the device has been properly CE marked and that the company from which they are importing the device has an EU DoC drawn up. In addition, the manufacturer and authorized representative of the device must be identifiable. The device should also be labeled appropriately as per the regulation requirements, and an instructions for use (IFU) document must be available. In addition, if applicable, a UDI should be assigned. If, for some reason or another, the importer has reason to believe that the device is not in conformance with the EU regulation, then that device should not be placed on the market.

Importers are also required to indicate on the device, or on the packaging of the device, or in a separate document their relevant details. They should also be able to verify that the device is registered in the electronic system as per Article 29 of the regulation. Their details should be added accordingly. It is the responsibility of the importer to ensure that storage or transport conditions do not jeopardize device compliance with the general safety and performance requirements (GSPR).

As per the new MDR, importers are also required to keep a register that tracks complaints received, non-conforming devices and of recalls and withdrawals. These should be available to the manufacturer, authorized representative and distributors with any relevant information as requested to allow for investigation. An extract of the requirements for importers from the MDR is provided below:

Article 13: General Obligations of Importers

1. Importers shall place on the Union market only devices that are in conformity with this Regulation.

2. In order to place a device on the market, importers shall verify that:

i. the device has been CE marked and that the EU declaration of conformity of the device has been drawn up;

ii. a manufacturer is identified and that an authorized representative in accordance with Article 11 has been designated by the manufacturer;

iii. the device is labeled in accordance with this Regulation and accompanied by the required instructions for use;

iv. where applicable, a UDI has been assigned by the manufacturer in accordance with Article 27.

Where an importer considers or has reason to believe that a device is not in conformity with the requirements of this Regulation, it shall not place the device on the market until it has been brought into conformity and shall inform the manufacturer and the manufacturer's authorized representative. Where the importer considers or has reason to believe that the device presents a serious risk or is a falsified device, it shall also inform the competent authority of the Member State in which the importer is established.

3. Importers shall indicate on the device or on its packaging or in a document accompanying the device their name, registered trade name or registered trademark, their registered place of business and the address at which they can be contacted, so that their location can be established. They shall ensure that any additional label does not obscure any information on the label provided by the manufacturer.

4. Importers shall verify that the device is registered in the electronic system in accordance with Article 29. Importers shall add their details to the registration in accordance with Article 31.

5. Importers shall ensure that, while a device is under their responsibility, storage or transport conditions do not jeopardize its compliance with the general safety and performance requirements set out in Annex I and shall comply with the conditions set by the manufacturer, where available.

6. Importers shall keep a register of complaints, of non-conforming devices and of recalls and withdrawals, and provide the manufacturer, authorized representative and distributors with any information requested by them, in order to allow them to investigate complaints.

7. Importers who consider or have reason to believe that a device which they have placed on the market is not in conformity with this Regulation shall immediately inform the manufacturer and its authorized representative. Importers shall co-operate with the manufacturer, the manufacturer's authorized representative, and the competent authorities to ensure that the necessary corrective action to bring that device into conformity, to withdraw or recall it is taken. Where the device presents a serious risk, they shall also immediately inform the competent authorities of the Member States in which they made the device available and, if applicable, the notified body that issued a certificate in accordance with Article 56 for the device in question, giving details, in particular, of the non-compliance and of any corrective action taken.

8.	Importers who have received complaints or reports from healthcare professionals, patients or users about suspected incidents related to a device which they have placed on the market shall immediately forward this information to the manufacturer and its authorized representative.
9.	Importers shall, for the period referred to in Article 10(8), keep a copy of the EU declaration of conformity and, if applicable, a copy of any relevant certificate, including any amendments and supplements, issued in accordance with Article 56.
10.	Importers shall cooperate with competent authorities, at the latter's request, on any action taken to eliminate or, if that is not possible, mitigate the risks posed by devices which they have placed on the market. Importers, upon request by a competent authority of the Member State in which the importer has its registered place of business, shall provide samples of the device free of charge or, where that is impracticable, grant access to the device.

10.7. GENERAL OBLIGATIONS OF DISTRIBUTORS AS ECONOMIC OPERATORS

General obligations have been set out in the MDR if a company serves as a distributor of a medical device. These can be found in Article 14 of the MDR. The regulation stipulates that before a distributor can make a device available on the market, they should verify if the device has been CE marked and that a suitable EU DoC is drawn up. In addition, distributors should also verify that the device is accompanied by the relevant information that should be supplied by the manufacturer as per requirements laid out in Article 10. Distributors should also ensure that the requirements laid down for importers are met, if in certain instances, the device itself is imported. Verification of an assignment of a UDI where applicable should also be done. If, for some reason or another, the distributor has reason to believe that the device is not in conformance with the requirements of the MDR, then that device should not be made available on the market until such time conformance can be proved. Communication to the manufacturer and related authorized representatives (AR) should also be made indicating non-conformance. If for some reason the distributor has reason to believe that the device may pose serious risk or is a falsified device, then communication should be made to the CA of the Member State in which it is established. It is the responsibility of the distributor to ensure that while the device is under their responsibility, storage or transport conditions comply with that defined by the manufacturer. In instances where corrective action is needed to bring the device into conformance, distributors should work together and comply with the manufacturer where possible. It is also the responsibility of distributors to forward any complaints or reports received from healthcare professionals, patients or users regarding the device to the manufacturer. Distributors are now required to keep a register of complaints, non-conforming devices and of recalls and withdrawals, which shall be made available to the manufacturer upon request.

An extract of the requirements for distributors from the MDR is provided below:

Article 14: General Obligations of Distributors

1. When making a device available on the market, distributors shall, in the context of their activities, act with due care in relation to the requirements applicable.
2. Before making a device available on the market, distributors shall verify that all of the following requirements are met:

i. the device has been CE marked and that the EU declaration of conformity of the device has been drawn up;

ii. the device is accompanied by the information to be supplied by the manufacturer in accordance with Article 10(11);

iii. for imported devices, the importer has complied with the requirements set out in Article 13(3);

iv. that, where applicable, a UDI has been assigned by the manufacturer.

In order to meet the requirements referred to in points (a), (b) and (d) of the first subparagraph, the distributor may apply a sampling method that is representative of the devices supplied by that distributor.

Where a distributor considers or has reason to believe that a device is not in conformity with the requirements of this Regulation, it shall not make the device available on the market until it has been brought into conformity, and shall inform the manufacturer and, where applicable, the manufacturer's authorized representative, and the importer. Where the distributor considers or has reason to believe that the device presents a serious risk or is a falsified device, it shall also inform the competent authority of the Member State in which it is established.

3. Distributors shall ensure that, while the device is under their responsibility, storage or transport conditions comply with the conditions set by the manufacturer.
4. Distributors that consider or have reason to believe that a device which they have made available on the market is not in conformity with this Regulation shall immediately inform the manufacturer and, where applicable, the manufacturer's authorized representative and the importer. Distributors shall co-operate with the manufacturer and, where applicable, the manufacturer's authorized representative, and the importer, and with competent authorities to ensure that the necessary corrective action to bring that device into conformity, to withdraw or to recall it, as appropriate, is taken. Where the distributor considers or has reason to believe that the device presents a serious risk, it shall also immediately inform the competent authorities of the Member States in which it made the device available, giving details, in particular, of the non-compliance and of any corrective action taken.

5. Distributors that have received complaints or reports from healthcare professionals, patients or users about suspected incidents related to a device they have made available, shall immediately forward this information to the manufacturer and, where applicable, the manufacturer's authorized representative and the importer. They shall keep a register of complaints, of non-conforming devices and of recalls and withdrawals, and keep the manufacturer and, where available, the authorized representative and the importer informed of such monitoring and provide them with any information upon their request.
6. Distributors shall, upon request by a competent authority, provide it with all the information and documentation that is at their disposal and is necessary to demonstrate the conformity of a device.

Distributors shall be considered to have fulfilled the obligation referred to in the first subparagraph when the manufacturer or, where applicable, the authorized representative for the device in question provides the required information. Distributors shall cooperate with competent authorities, at their request, on any action taken to eliminate the risks posed by devices which they have made available on the market. Distributors, upon request by a competent authority, shall provide free samples of the device or, where that is impracticable, grant access to the device.

10.8. GENERAL OBLIGATIONS OF AUTHORIZED REPRESENTATIVES (AR) AS ECONOMIC OPERATORS

The European Authorized Representative serves as the liaison between the manufacturer and the national competent authorities or Ministries of Health in the EU. A European Authorized Representative may also be known as the EUAR, EC Rep, CE Rep, EU Rep and/or an EAR. All of the abovementioned abbreviations mean the same thing: European Authorized Representative. The European Authorized Representative serves as the point of contact between the manufacturer (which is a non-European medical device company) and the National Competent Authorities and notified bodies (NB).

The roles and responsibilities of the Authorized Representative include:

- Operating in accordance with MEDDEV 2.5/10 Guidelines for AR.
- Assisting with medical device registrations within the European Union (EU)
- Holding a copy of the manufacturer's technical file and/or CE DoC in case it is requested by the relevant National Competent Authority.

- Assisting with Incident and field safety corrective actions (FSCA) reporting as per MEDDEV 2.12/1
- Provide, if requested, a regulatory review of product artwork and labeling.
- Provide regulatory news and updates.

The name and address of the Authorized Representative must be included on the labeling, outer packaging and IFU and the appropriate EC Rep symbol must also be used as per EN ISO 15223-1:2016.

It is possible to also appoint a distributor as an Authorized Representative in Europe; however, this is usually not recommended because most distributors are not prepared to fulfill the mandatory responsibilities of the EC Representative. Manufacturers should instead opt to select an independent representative that focuses directly on regulatory affairs.

It is also important to note that under the new MDR 2017/745, AR will have greater responsibilities and will take on significantly more risk and liabilities. These responsibilities are often laid out in the agreement between an Authorized Representative and the manufacturer.

An extract from the MDR on AR is given below:

Article 11: Authorized representative

1. Where the manufacturer of a device is not established in a Member State, the device may only be placed on the Union market if the manufacturer designates a sole authorized representative.
2. The designation shall constitute the authorized representative's mandate; it shall be valid only when accepted in writing by the authorized representative and shall be effective at least for all devices of the same generic device group.
3. The authorized representative shall perform the tasks specified in the mandate agreed between it and the manufacturer. The authorized representative shall provide a copy of the mandate to the competent authority, upon request.

The mandate shall require, and the manufacturer shall enable, the authorized representative to perform at least the following tasks in relation to the devices that it covers:

i. Verify that the EU declaration of conformity and technical documentation have been drawn up and, where applicable, that an appropriate conformity assessment procedure has been carried out by the manufacturer;

ii. keep available a copy of the technical documentation, the EU declaration of conformity and, if applicable, a copy of the relevant certificate, including any amendments and supplements, issued in accordance with Article 56, at the disposal of competent authorities for the period referred to in Article 10(8);

iii.	comply with the registration obligations laid down in Article 31 and verify that the manufacturer has complied with the registration obligations laid down in Articles 27 and 29;
iv.	in response to a request from a competent authority, provide that competent authority with all the information and documentation necessary to demonstrate the conformity of a device, in an official Union language determined by the Member State concerned;
v.	forward to the manufacturer any request by a competent authority of the Member State in which the authorized representative has its registered place of business for samples, or access to a device and verify that the competent authority receives the samples or is given access to the device;
vi.	cooperate with the competent authorities on any preventive or corrective action taken to eliminate or, if that is not possible, mitigate the risks posed by devices;
vii.	immediately inform the manufacturer about complaints and reports from healthcare professionals, patients, and users about suspected incidents related to a device for which they have been designated;
viii.	terminate the mandate if the manufacturer acts contrary to its obligations under this Regulation.
4.	The mandate referred to in paragraph 3 of this Article shall not delegate the manufacturer's obligations laid down in Article 10(1), (2), (3), (4), (6), (7), (9), (10), (11) and (12).
5.	Without prejudice to paragraph 4 of this Article, where the manufacturer is not established in a Member State and has not complied with the obligations laid down in Article 10, the authorized representative shall be legally liable for defective devices on the same basis as, and jointly and severally with, the manufacturer.
6.	An authorized representative who terminates its mandate on the ground referred to in point (h) of paragraph 3 shall immediately inform the competent authority of the Member State in which it is established and, where applicable, the notified body that was involved in the conformity assessment for the device of the termination of the mandate and the reasons therefor.
7.	Any reference in this Regulation to the competent authority of the Member State in which the manufacturer has its registered place of business shall be understood as a reference to the competent authority of the Member State in which the authorized representative, designated by a manufacturer referred to in paragraph 1, has its registered place of business.

The regulation also stipulates what a manufacturer should do when there is a change in the Authorized Representative:

Article 12: Change of Authorized Representative
The detailed arrangements for a change of authorized representative shall be clearly defined in an agreement between the manufacturer, where practicable the outgoing authorized representative, and the incoming authorized representative. That agreement shall address at least the following aspects: i. the date of termination of the mandate of the outgoing authorized representative and date of beginning of the mandate of the incoming authorized representative; ii. the date until which the outgoing authorized representative may be indicated in the information supplied by the manufacturer, including any promotional material; iii. the transfer of documents, including confidentiality aspects and property rights; iv. the obligation of the outgoing authorized representative after the end of the mandate to forward to the manufacturer or incoming authorized representative any complaints or reports from healthcare professionals, patients or users about suspected incidents related to a device for which it had been designated as authorized representative.

Chapter 11

Designating a Person Responsible for Regulatory Compliance

CONTENTS

11.1. PERSON RESPONSIBLE FOR REGULATORY COMPLIANCE (PRRC)

The new MDR requires medical device companies to appoint a person that will be responsible for regulatory compliance, also known as the PRRC (person responsible for regulatory compliance). The PRRC is responsible for the technical documentation, ensuring the EU DoC is drawn up and kept up to date, post market surveillance and checking conformity of devices before they are released as per the quality management system (QMS) that has been implemented. In addition, the PRRC is also responsible for carrying out vigilance activities, recording, and reporting incidents, safety, and performance requirements, and clinical investigations. It is also possible for companies to have multiple PRRC's which carry out different regulatory functions as long their roles and responsibilities are recorded. In addition, manufacturers will need to submit the name, address, and contact details of their PRCC(s) upon registration and this should be kept up to date at all times.

11.1.1. Qualifications of the PRRC

The EU MDR (as well as the new IVDR) further defines what the related qualifications of the PRRC should be. As per Article 15 of the MDR, these include:

- A diploma, certificate or other evidence of formal qualification in in law, medicine, pharmacy, engineering or another relevant scientific discipline;
- At least one year of professional experience in regulatory affairs or in QMSs relating to medical devices; or
- Four years of professional experience in regulatory affairs or in QMSs relating to medical devices.

11.1.2. Does Every Medical Device Manufacturer Require a PRRC?

Appointment of a PRRC depends on the size of the organization. Small manufacturers with fewer than 50 full-time employees globally and under 10 million Euros in global sales revenue (not just the EU) are not required to have a PRRC within the company. These small manufacturers can outsource the role but are still required to have such a regulatory individual at their disposal.

Article 15: Person Responsible for Regulatory Compliance

1. Manufacturers shall have available within their organization at least one person responsible for regulatory compliance who possesses the requisite expertise in the field of medical devices. The requisite expertise shall be demonstrated by either of the following qualifications:

i. a diploma, certificate or other evidence of formal qualification, awarded on completion of a university degree or of a course of study recognized as equivalent by the Member State concerned, in law, medicine, pharmacy, engineering or another relevant scientific discipline, and at least one year of professional experience in regulatory affairs or in quality management systems relating to medical devices;

ii. four years of professional experience in regulatory affairs or in quality management systems relating to medical devices.

Without prejudice to national provisions regarding professional qualifications, manufacturers of custom-made devices may demonstrate the requisite expertise referred to in the first subparagraph by having at least two years of professional experience within a relevant field of manufacturing.

2. Micro and small enterprises within the meaning of Commission Recommendation 2003/361/EC (1) shall not be required to have the person responsible for regulatory compliance within their organization but shall have such person permanently and continuously at their disposal.

3. The person responsible for regulatory compliance shall at least be responsible for ensuring that:

i. the conformity of the devices is appropriately checked, in accordance with the quality management system under which the devices are manufactured, before a device is released;

ii. the technical documentation and the EU declaration of conformity are drawn up and kept up-to-date;

iii. the post-market surveillance obligations are complied with in accordance with Article 10(10);

iv. the reporting obligations referred to in Articles 87 to 91 are fulfilled;

v. in the case of investigational devices, the statement referred to in Section 4.1 of Chapter II of Annex XV is issued.

4. If a number of persons are jointly responsible for regulatory compliance in accordance with paragraphs 1, 2 and 3, their respective areas of responsibility shall be stipulated in writing.

5. The person responsible for regulatory compliance shall suffer no disadvantage within the manufacturer's organization in relation to the proper fulfillment of his or her duties, regardless of whether or not they are employees of the organization.

6. Authorized representatives shall have permanently and continuously at their disposal at least one person responsible for regulatory compliance who possesses the requisite expertise regarding the regulatory requirements for medical devices in the Union. The requisite expertise shall be demonstrated by either of the following qualifications:

i.	a diploma, certificate or other evidence of formal qualification, awarded on completion of a university degree or of a course of study recognized as equivalent by the Member State concerned, in law, medicine, pharmacy, engineering or another relevant scientific discipline, and at least one year of professional experience in regulatory affairs or in quality management systems relating to medical devices;
ii.	four years of professional experience in regulatory affairs or in quality management systems relating to medical devices.

The Medical Device Coordination Group (MDCG) established by Article 103 of the MDR 2017/745 has since released MDCG-2019-7 which a Guidance document on Article 15 of the MDR regarding a PRRC. Further explanation from MDCG 2019-7 is provided below.

As per Article 15: Manufacturers (Paragraph 1)

"Manufacturers shall have available within their organization at least one person responsible for regulatory compliance who possesses the requisite expertise in the field of medical devices. The requisite expertise shall be demonstrated by either of the following qualifications:
i. a diploma, certificate or other evidence of formal qualification, awarded on completion of a university degree or of a course of study recognized as equivalent by the Member State concerned, in law, medicine, pharmacy, engineering or another relevant scientific discipline, and at least one year of professional experience in regulatory affairs or in quality management systems relating to medical devices;
ii. four years of professional experience in regulatory affairs or in quality management systems relating to medical devices.
Without prejudice to national provisions regarding professional qualifications, manufacturers of custom-made devices may demonstrate the requisite expertise referred to in the first subparagraph by having at least two years of professional experience within a relevant field of manufacturing."

The guidance document provides clarification on the qualifications of the PRRC. It should be noted that any qualification acquired outside of the EU, including any university diplomas or certificates should have been recognized by an EU Member State as equivalent to the EU corresponding qualification. In addition, any professional experience in regulatory affairs or in QMSs relating to medical devices should be related to the EU requirements in the field.

The guidance document further explains the meaning of *"within their organization"* – the PRRC that is appointed would need to be an actual employee of the organization. In addition, organizations with more than one legal manufacturer under the parent company would need to ensure that each legal manufacturer has its own PRRC. The location of the PRRC is also important as a close link between the PRRC and the manufacturer should exist. Therefore, for those manufacturers located outside the EU, it must be assumed that the PRRC should also be located outside the EU, whilst for manufacturers located in the EU, it must be assumed that the PRRC should also be located in the EU.

As per Article 15: Micro and Small Manufacturers (Paragraph 2)

"Micro and small enterprises within the meaning of Commission Recommendation 2003/361/EC shall not be required to have the person responsible for regulatory compliance within their organization but shall have such person permanently and continuously at their disposal."

Paragraph 2 of Article 15 in the Medical Device Directive refers to micro and small enterprises having a PRRC "permanently and continuously at their disposal." This means that the micro or small enterprise may subcontract the responsibilities of a PRRC to a third party such a regulatory consultant, as long as the qualification criteria is met and the manufacturer can demonstrate and document how they can meet their legal obligations. A common example is that the PRRC may be part of an external organization with which the manufacturer has established a contract with defined roles and responsibilities and whereby the manufacturer has ensured the permanent and continuous availability of the party. The contract should also mention the relevant persons qualifications allowing compliance with the regulation.

It is also important to note that for micro and/or small enterprises located in the EU, it must be assumed that any person that shall be "permanently and continuously at their disposal" should also be located in the EU.

As per Article 15: Authorized Representatives (Paragraph 6)

"Authorized representatives shall have permanently and continuously at their disposal at least one person responsible for regulatory compliance who possesses the requisite expertise regarding the regulatory requirements for medical devices in the Union. The requisite expertise shall be demonstrated by either of the following qualifications:

i. a diploma, certificate or other evidence of formal qualification, awarded on completion of a university degree or of a course of study recognized as equivalent by the Member State concerned, in law, medicine, pharmacy, engineering or another relevant scientific discipline, and at least one year of professional experience in regulatory affairs or in quality management systems relating to medical devices;

ii. four years of professional experience in regulatory affairs or in quality management systems relating to medical devices."

Paragraph 6 of Article 15 in the Medical Device Directive refers to the requirements for authorized representatives (AR), who are also required to have a PRRC "permanently and continuously at their disposal." The authorized representative may subcontract the responsibilities of a PRRC to a third party, so long as the qualification criteria is met and the authorized representative can demonstrate and document how they can meet their legal obligations. An example of this is that the PRRC may be part of an external organization with which the authorized representative has established a contract laying down the requirements so as ensure the permanent and continuous availability of the PRRC. The contract should also mention the relevant PRRC's qualifications showing compliance with the regulation.

As the Authorized Representative is located in the EU, it must be assumed that any person to be permanently and continuously at their disposal shall be located in the EU as well.

As per Article 15: Roles and Responsibilities of the Person Responsible for Regulatory Compliance within a Manufacturer (Paragraph 3)

The person responsible for regulatory compliance shall at least be responsible for ensuring that:

a. The conformity of the devices is appropriately checked, in accordance with the quality management system under which the devices are manufactured, before a device is released.

The roles and responsibilities of a PRRC are often cross-referenced to the roles and responsibilities of a manufacturer, which is stated in Article 10 of the MDR. In addition, any guidance on post-market surveillance, vigilance, clinical investigations, and related performance studies should be incorporated when determining the roles and responsibilities.

Manufacturers of devices, other than investigational devices, shall establish, document, implement, maintain, keep up to date and continually improve a QMS that shall ensure compliance with the Regulation in the most effective manner and in a manner that is proportionate to the risk class and the type of device.

As per Article 15: Roles and Responsibilities of the Person Responsible for Regulatory Compliance within a Manufacturer (Paragraph 3)

b. the technical documentation and the EU declaration of conformity are drawn up and kept up-to-date.

Manufacturers of medical devices other than custom-made devices shall draw up and keep up-to-date technical documentation for those devices as well as an EU DoC.

As per Article 15: Roles and Responsibilities of the Person Responsible for Regulatory Compliance within a Manufacturer (Paragraph 3)

c. the post-market surveillance obligations are complied with in accordance with Article 10(10) [Article 10(9) of the IVDR].

Manufacturers of devices shall implement and keep up to date the post-market surveillance system.

As per Article 15: Roles and Responsibilities of the Person Responsible for Regulatory Compliance within a Manufacturer (Paragraph 3)

d. the reporting obligations referred to in Articles 87 to 91 [Article 82 and 86 of the IVDR] are fulfilled.

Manufacturers shall have a system for recording and reporting of incidents and FSCA. This is described in Article 10(13) of the MDR.

As per Article 15: Roles and Responsibilities of the Person Responsible for Regulatory Compliance within a Manufacturer (Paragraph 3)

e. in the case of investigational devices, the statement referred to in Section 4.1 of Chapter II of Annex XV [Section 4.1 of Annex XIV of the IVDR] is issued.

Manufacturers shall ensure that "a signed statement by the natural or legal person responsible for the manufacture of the investigational device that the device in question conforms to the general safety and performance requirements (GSPR) apart from the aspects covered by the clinical investigation and that, with regard to those aspects, every precaution has been taken to protect the health and safety of the subject."

11.1.3. Can One Person Be the PRRC for a Manufacturer and Its Authorized Representative?

The PRRC for an authorized representative cannot be the same person for an 'outside EU' manufacturer. Within the regulation, it is clear that the authorized representative should be adding an additional level of scrutiny to the device and to ensure that the supervision and control of the manufacture of devices, and the relevant post-market surveillance and vigilance activities are adequately carried out. Therefore, if two roles were conducted by the same person, the additional level of scrutiny would be undermined. For the same reason, the PRRC of a micro or small enterprise and the PRRC of the authorized representative of that same enterprise shall not belong to the same external organization.

11.2. CONCLUSION

The need for improving regulatory controls and finding ways to improve the quality and access to safe medical devices has increased significantly over the last few years. For centuries, society has benefited from a host of discoveries and developments that have centered around improving health care. The use of medical technology, however, dates back to the first half of the 19th century. While this reaches far back, evidence suggests the actual application of medical devices and modern medical technology only took off over the last 50 years. As a result, medical devices have since become an essential part of health care and a vital component of the various activities carried out by healthcare providers to treat people with various medical conditions.

As science and technology continue to advance, so research continues to be conducted in this field, leading to extensive progress within the development of various medical technologies. Included in these are genetic testing, genetic manipulations and remote patient management. Despite this progress, most of the world's population continues to have little or no access to much of these advances, often due to a lack of a suitable regulatory framework. The establishment of an effective regulatory system is dependent on numerous factors including long-term government commitment and an overall political and societal willingness to cooperate. As you would have now seen, the new MDR (2017/745) is not entirely different from the medical device directive (MDD) as many of the related standards have been continuously updated in the last 25 years since the MDD existed. For this reason, much of the changes in the new regulation are essentially a more stringent or stricter form of previous requirements. Changes brought about by the MDR in Europe require not only manufacturers but all defined economic operators to take a proactive, multilevel, and multidisciplinary approach when it comes to ensuring safe and effective devices are placed on the market. Impacts will be felt on all levels of the business, from supply chain to quality to marketing and distribution. For this reason, ensuring a thorough understanding of the MDR as well as being able to properly identify its full impact on each section of respective businesses is critical to a successful and effective transition.

Bibliography

1. *A Disaster Relief Supply Management System (SUMA)*. Pan American Health Organization/WHO Regional Office for the Americas, Washington, 1996.
2. Ahrq.gov. *Agency for Health Research and Quality*; Brockton Heflin; Center for Devices and Radiological Health; Food and Drug Administration. Available from: http//www.healthit.ahrq.gov/portal/server.pt?open=18.ppt (accessed on 3 April 2021).
3. Ahwp.info. *Asian Harmonization Working Party*. Report from AHWP STG Nomenclature. Available from: http://www.ahwp.info/ClientFolder/AHWP/Library/Tree/4_Technical_Committee/3_TC_Minutes/The_9th_AHWP_Technical_Committee_Meeting/14_Medical_Device_Nomenclautre.pdf (accessed on 3 April 2021).
4. American Medical Association, (2003). World medical association declaration of Helsinki. Ethical principles for medical research involving human subjects. *Clinical Review and Education*. 2191–2194.
5. *Application of Risk Management to Medical Devices*, (2007). International Organization for Standardization (ISO 14971:2007 Standard). https://www.iso.org/obp/ui/#iso:std:iso:14971:ed-2:v2:en (accessed on 3 April 2021).
6. Bloom, G., & Temple-Bird, C., (1990). *Medical Equipment in Sub-Saharan Africa: A Framework for Policy-Formulation.* IDS Research Report No. 19, WHO/SHS/NHP/90.6, WHO, Geneva.
7. Carlson, R. V., Boyd, K. M., & Webb, D. J., (2004). The revision of the

Declaration of Helsinki: past, present, and future. *British Journal of Clinical Pharmacology, 57*, 695–713.

8. Chang, T., (1997). Medical device donations to developing countries: Things to consider. *Poster Presentation at the 32nd AAMI Annual Meeting*. Washington, D.C., USA.
9. *Clinical Investigations*, (2010). Global harmonization task force (GHTF/SG5/ N3:2010). http://www.imdrf.org/docs/ghtf/final/sg5/technical-docs/ghtf-sg5-n3-clinical-investigations-100212.pdf (accessed on 3 April 2021).
10. *Development of Medical Device Policies*, (2011). Geneva: World Health Organization. WHO Medical Device Technical Series. http://apps.who.int/medicinedocs/documents/s21559en/s21559en.pdf (accessed on 3 April 2021).
11. *Emergency Relief Items: Compendium of Basic Specifications: Medical Supplies and Equipment*, (1996). Vol. 2. UNDP/WHO, Geneva.
12. EU Council Directive 93/42/EEC, (1993). Concerning medical devices http://www.imdrf.org/docs/ghtf/archived/sg4/technical-docs/ghtf-sg4-guidelinesauditing-qms-part-1-general-requirements-080827.pdf (accessed on 3 April 2021).
13. *Executive Report of the Regional Workshop on Health care Technology in Sub-Saharan Region, Somerset West, South Africa*. South African Medical Research Council, Tygerberg, 1994.
14. Facinelli, G., (1994). Biomedical technology and clinical engineering in cooperation programs with developing countries: Methodological guidelines. In: *Proceedings of the Regional Workshop on Health Care Technology in the Sub-Saharan Region, Somerset West, South Africa* (p. 13). South African Medical Research Council, Tygerberg.
15. Ghtf.org. *Global Harmonization Task Force; Generating the GMDN Process and Development*. Available from: http//www.ghtf.org/meetings/conferences/8thconference/../Dillard.ppt (accessed on 3 April 2021).
16. *Global Medical Devices Nomenclature System (GMDN)* [website] https://www.gmdnagency.org (accessed on 3 April 2021).
17. Gmdnagency.com. *Global Medical Device Nomenclature*. Available from: http://www.gmdnagency.com/?id=userinfo (accessed on 3 April 2021).
18. *Good Governance for Medicines: Model Framework*, (2014).

Geneva: World Health Organization. http://apps.who.int/iris/bitstream/10665/129495/1/9789241507516_eng.pdf?ua=1&ua=1 (accessed on 3 April 2021).

19. *Guidelines for Donors and Recipients of Equipment Donations*, (1990). Joint Medical Stores, Kampala, Uganda.
20. *Guidelines for Donors and Recipients of Pharmaceutical Donations*, (1990). Second Revised Edition. Christian Medical Commission, World Council of Churches, Geneva.
21. *Guidelines for Drug Donations*, (1996). WHO/DAP/96.2, WHO, Geneva.
22. *Guidelines for Medical Equipment Donation*, (1995). American College of Clinical Engineering, Plymouth Meeting, PA.
23. *Guidelines for Purchasing of Medical Equipment*, (1991). Presbyterian Technical Services, Presbyterian Church in Cameroon, Kumba, Cameroon.
24. *Guidelines on Donations of Drugs and Medical Equipment to the Health Sector for Tanzania Mainland*, (1995). Ministry of Health, Dar-es-Salaam, Tanzania.
25. *Guidelines: Medical Equipment Donations*, (1994). Technical Library, FAKT, in: CMC Publication CONTACT 10/94. CMC, World Council of Churches, Geneva.
26. *Guiding Principles for Regulatory Quality and Performance*, (2005). Paris: Organization for Economic Cooperation and Development (OECD). http://www.oecd.org/fr/reformereg/34976533.pdf (accessed on 3 April 2021).
27. Heimann, P., & Poluta, M. A., (1997). *Health Technology Management in the Sub-Saharan Region as a Prerequisite for Optimizing the Donor Aid Intervention Process* (In press) WHO, ARA, Geneva.
28. IMDG, (1992). *Donating and Selling Used Medical Equipment* (Vol. 21, No. 9, pp. 295–297). Health Devices, ECRI, Plymouth Meeting, PA.
29. *International Health Relief Assistance: A Guide for Effective Aid.* Pan American Health Organization/WHO Regional Office for the Americas, Washington, 1990.
30. Issakov, A., (1994). Service and maintenance in developing countries. In: Van, G. C. W. D., (ed.), *Medical Devices: International Perspectives*

on Health and Safety (pp. 21–28). Elsevier, Amsterdam.

31. Kachienga, M. O., (1992). *An Overview: Procurement of Health Care Equipment for Public Health Facilities* (pp. 20–23). Medicus.
32. Madani, M. A., (1995). Medical equipment retirement and donation criteria. *Presentation to the 30th AAMI Annual Meeting*. Anaheim, CA, USA.
33. *Medical Device Regulations: Global Overview and Guiding Principles*, (2003). Geneva: World Health Organization. http://www.who.int/medical_devices/publications/en/MD_Regulations.pdf (accessed on 3 April 2021).
34. MHRA (Medicines and Healthcare products Regulatory Agency), (2010a). *Medical Device Alert: DePuy ASR Acetabular Cups Used in Hip Resurfacing Arthroplasty and Total Hip Replacement*. MDA/2010/044.
35. *Policy on Acquiring Medical Equipment Through Quotation/Tender and Donation (Loose Minute)*, (1987). Ministry of Health, Nairobi, Kenya.
36. Prage, L., (1987). *Guidelines for Support to Procurement, Operation, and Maintenance of Scientific Equipment in Developing Countries*. IFS/SAREC, Stockholm, Sweden.
37. *Principles of Conformity Assessment for Medical Devices*, (2012). *Global Harmonization Task Force (GHTF/SG1/N78:2012)*. http://www.imdrf.org/docs/ghtf/final/sg1/technical-docs/ghtf-sg1-n78-2012-conformity-assessment-medicaldevices-121102.pdf (accessed on 3 April 2021).
38. *Principles of Medical Devices Classification*, (2012). *Global Harmonization Task Force (GHTF/SG1/N77:2012*). http://www.imdrf.org/docs/ghtf/final/sg1/technicaldocs/ghtf-sg1-n77-2012-principles-medical-devices-classification-121102.pdf (accessed on 3 April 2021).
39. *Quality Management System-Medical Devices-Guidance on Corrective Action and Preventive Action and Related QMS Processes*, (2010). *Global Harmonization Task Force (GHTF/SG3/N18:2010)*. http://www.imdrf.org/docs/ghtf/final/sg3/technical-docs/ghtf-sg3-n18-2010-qms-guidance-on-corrective-preventativeaction-101104.pdf (accessed on 3 April 2021).
40. *Regulatory Framework for Control of Refurbished Medical Devices*, (2012). *Asia Pacific Economic Cooperation*. http://mddb.apec.org/

Documents/2012/MAG/WKSP1/12_mag_wksp1_011.pdf (accessed on 3 April 2021).

41. *Regulatory Harmonization and Convergence*. Silver Spring (MD): US Food and Drug Administration (US FDA). http://www.fda.gov/BiologicsBloodVaccines/InternationalActivities/ucm271079.htm (accessed on 3 April 2021).
42. *Report of the Interregional Meeting on Maintenance and Repair of Health Care Equipment, Nicosia, Cyprus*, (1987). WHO/SHS/NHP/87.5, WHO, Geneva.
43. *Role of Standards in the Assessment of Medical Devices* (2008). Global Harmonization Task Force; (GHTF/SG1/N044:2008). http://www.imdrf.org/docs/ghtf/final/sg1/procedural-docs/ghtf-sg1-n044-2008-standards-in-assessment-of-medicaldevices-080305.pdf (accessed on 3 April 2021).
44. Rosenblatt, W. H., (1997). *Surplus Medical Supplies*. To be published in World Health Magazine.
45. *Safety Aspects-Guidelines for Their Inclusion in Standards*, (2014). Geneva: International Organization for Standardization; (ISO/IEC Guide 51:2014). http://www.iso.org/iso/catalogue_detail.htm?csnumber=53940 (accessed on 3 April 2021).
46. Study on Drugs, Equipment, and Supplies and the Role of Christian Medical Association of India, (1987). CMAI, New Delhi.
47. Wang, B., (1996). The facts on used equipment export. *Biomedical Technology Management*, 64.
48. *WHO Global Governance*. http://www.who.int/healthpromotion/conferences/9gchp/good-governance/en/ (accessed on 3 April 2021).
49. *WHO Good Regulatory Practices: Guidelines for National Regulatory Authorities for Medical Products*. Geneva: World Health Organization (under development, 2017).
50. World Health Organization, (2001). *Bulletin of the World Health Organization*. World Medical Association Declaration of Helsinki, Ethical Principles for Medical Research Involving Human Subjects.
51. World Medical Association Declaration of Helsinki: Ethical Principles for Medical Research Involving Human Subjects, (2013). American medical association. *JAMA, 310*, 2191–2194.

INDEX

H

I

J

L

M

N

U

V

W